The cattle sector in Central and Eastern Europe

The cattle sector in Central and Eastern Europe

Developments and opportunities in a time of transition

Editors:

K.J. Peters

A. Kuipers

M.G. Keane

A. Dimitriadou

EAAP Technical Series No. 10

ISBN 978-90-8686-104-0
ISSN 1570-7318

First published, 2009

© Wageningen Academic Publishers
The Netherlands, 2009

Table of contents

Part 3 Concluding remarks

Introduction

The countries in Eastern-Europe are in a lengthy period of rapid change. Ten Central and Eastern European countries entered the European Union in 2004 and two entered in 2007. Surrounding countries to the East are in a similar process of change following the disintegration of the former Soviet Union. The Cattle Network of the EAAP has actively analysed the transformation processes in a number of EAAP annual meetings. Presentations are available on the web page of the Cattle Network. In 2004, as part of the EAAP meeting in Bled, Slovenia, a workshop was organised dealing with the consequences of EU entry for a group of countries. The 'new' EU-countries, but also all neighbouring countries to the East, participated in this meeting. Results of that meeting are available in the EAAP Technical Series publication No. 8: 'Farm management and extension needs in Central and Eastern European countries under the EU milk quota'.

This workshop was organised by the Cattle Network Working Group, the Central and Eastern European Working Group, and the Cattle Commission of EAAP during the annual meeting of the EAAP in Vilnius, Lithuania. The Eastern location was one of the motives for doing so. The FAO again generously supported the workshop as before. About 60 persons from a wide variety of countries attended.

Ms. Andie Dimitriadou of the EAAP secretariat took care of the logistics of the meeting together with the travel arrangements of the invited speakers. She also collected and distributed the Power Point presentations. Later on, she collected the publications which are compiled in this book. We are very grateful for her support. The original idea for the workshop came from Mr. Kurt Peters and Mr. Abele Kuipers. We together with Mr. Gerry Keane reviewed the papers thoroughly. Mr. Arunas Svitojus, Baltic Heifer Foundation, helped as local organiser realising the workshop and in the selection of the invited speakers. Mr. Andriy Rozstalnyy, FAO Animal Production and Health Officer at Subregional office for Central and Eastern Europe was very supportive in the organisation of the workshop. Special gratitude is given to Mrs. Maria Kadlecikova FAO Regional Representative for Europe and Central Asia for making this workshop possible.

This book is comprised of 13 contributions: four overview articles, one on the topic of animal welfare and eight country reports. The country reports come from a wide variety of countries in Eastern Europe and Asia: Slovakia, Poland, Baltic States, Russian Federation, Belarus Ukraine, Caucasian countries (Georgia and Azerbaijan) and Central Asian countries (Kazakhstan, Kyrgyzstan and Uzbekistan). The country reports describe the transition taking place in these countries. Very positive was the fact that the developments in the beef cattle sector as well as in the dairy chain are described. The mix of participating speakers from universities, research and developmental institutions, farmers' organisations, agribusiness clubs and marketing boards was very beneficial. Some analyses were made and several critical points in development were signalled. Thus, barriers as well as opportunities for further development are mentioned and described in this book. The discussions during the workshop were of a high quality. Some conclusions are added as a last chapter in the book.

Kurt Peters and Abele Kuipers

Part 1 Overview articles

Beef sector challenges and perspectives in new EU member states

R. Csillag[1], A. Rozstalnyy[1], I. Hoffmann[2] and S. Mack[2]

[1]*Food and Agriculture Organization of the United Nations, Subregional Office for Central and Eastern Europe, Budapest, Hungary; andriy.rozstalnyy@fao.org;* [2]*Food and Agriculture Organization of the United Nations, Rome, Italy*

Abstract

Meat consumption declined during the 1990s due to the fall in purchasing power. Since 2000, meat consumption has been increasing and with the exceptions of Poland and Hungary, demand for meat cannot be supplied from domestic production. Many slaughter houses have closed but the remainder have been modernised to meet EU standards. Red meat is losing market share to poultry due to health scares, high price and changing dietary attitudes.

Keywords: beef, consumption trends, processing

Introduction

Since EU accession, the national bovine herd size has declined in most of the CEE countries with a corresponding fall in production and processing capacity (Table 1 and Table 2). Accession to the European Union (EU) has however stimulated greater intensification and concentration within the meat production and processing sub-sector. Yet farm structures remain fragmented with many small, family-owned, semi-subsistence units characterised by low efficiency and relatively high production costs. The slow pace of intensification and the move to more efficient production methods present a real barrier to the development of an efficient and integrated meat sector. More than two-thirds of beef, and virtually of all veal production, originate from the dairy herd. Sires from meat breeds are used to produce crossbreds to be reared for beef, but only one third of the beef derives from specific beef breeding herds. Beef production has fallen in recent years.

Table 1. Total of cattle population (×1,000 head).

	2000	2001	2002	2003	2004	2005	2006	2007
Bulgaria	652.2	641.1	699.0	736.2	679.6	630.0	636.5	611.0
Czech Republic	1,582.0	1,520.0	1,462.0	1,427.0	1,367.6	1,351.6	1,389.6	1,366.7
Estonia	252.8	260.5	253.9	257.2	249.8	252.2	245.0	242.0
Latvia	366.7	384.7	388.1	378.6	371.1	385.2	377.1	398.7
Lithuania	748.3	751.7	779.1	812.1	792.0	800.3	838.8	787.9
Hungary	805.0	783.0	770.0	739.0	723.0	708.0	702.0	705.0
Slovenia	493.7	477.1	473.2	449.9	451.1	452.5	454.0	479.6
Slovakia	646.1	625.2	607.8	593.2	540.1	527.9	507.8	501.8
Poland	5,723.0	5,498.8	5,421.0	5,276.8	5,200.2	5,385.0	5,281.0	5,405.5
Romania	2,870.4	2,799.8	2,877.8	2,897.1	2,808.1	2,861.1	2,933.6	2,819.0

Sources: FAOSTAT, EUROSTAT, Czech Statistical Office (CSÚ), Statistical Office of The Slovak Republic, Statistical Office of Estonia, Statistical Office of Lithuania and USDA.

Table 2. Number of bovines slaughtered (×1,000 head).

	2000	2001	2002	2003	2004	2005	2006	2007
Bulgaria	422.00						161.27	
Czech Republic	374.46	355.44	374.86	372.97	335.81	281.04	273.58	269.66
Estonia	106.40	83.10	87.80	72.50	84.45	66.75	69.91	70.76
Latvia	181.72	143.34	107.24	133.24	132.50	118.92	114.54	119.57
Lithuania	562.50	416.60	372.00	330.20	337.17	312.28	238.75	252.43
Hungary	173.50			148.22	145.45	121.92	126.00	130.19
Poland	2,069.00	1,931.00	1,665.00	1,873.00	1,314.90	1,290.91	1,463.09	1,537.96
Romania							1,115.00	1,206.00
Slovakia	130.23	140.39	150.75	162.32	151.03	137.62	136.78	128.09
Slovenia	119.32	99.08	93.80	112.40	98.85	99.67	81.71	87.45

Sources: FAOSTAT, EUROSTAT, Czech Statistical Office (CSÚ), Statistical Office of The Slovak Republic, Statistical Office of Estonia, Statistical Office of Lithuania and USDA.

Meat consumption

The trend in meat consumption is an important social indicator and it is not surprising that meat consumption fell significantly during the mid 1990's mirroring the collapse of incomes and purchasing power. Since 2000, however, total meat consumption has started to increase throughout the region (Table 3). Increasing incomes in the non-agricultural sector is raising the demand for meat that cannot be supplied by domestic production. Most of the new EU member states remain net importers of beef with the exception of Poland and Hungary.

Processing

The number of slaughterhouses and processing plants has also declined since 1990. The larger enterprises received considerable investment during the preparation for EU accession in order to modernise and harmonise their operation to meet EU standards. A significant proportion of the abattoirs were however small 'one-room' operations. Most of these have had to close due to lack of capital and skills necessary for meeting the strict EU sanitary and traceability requirements as well as animal welfare, environmental (handling high risk materials, waste and effluent disposal) and meat inspection regulations.

Table 3. Beef and veal consumption (kg/capita/year).

	2000	2001	2002	2003	2004	2005	2006	2007
Slovakia	9.3	7.0	6.8	6.9	6.4	6.9		
Hungary	4.3	3.9	4.3	4.1	3.9	3.1		
Poland	8.0	6.3	5.9	6.7		5.0	4.0	
Czech Republic	12.5		11.6	11.3	10.4	10.0		
Estonia		12.1	11.4	13.3	10.4	12.8	12.4	13.8
Lithuania	15.0	11.2	10.0	10.5	10.6	8.0		
Romania		4.7	3.5	3.4	4.4	4.9		
Bulgaria	5.0	4.9	4.9	5.1	5.1	5.1		

Sources: FAOSTAT, EUROSTAT, Czech Statistical Office (CSÚ), Statistical Office of the Slovak Republic, Statistical Office of Estonia, Statistical Office of Lithuania and USDA.

Decline in red meat consumption

Consumption patterns have also changed in the past decade. Red meat consumption (including beef) has declined while poultry consumption has steadily increased with the exception of a temporary dip resulting from the avian influenza scare. The changing consumer preference away from red meat originates from the outbreak of BSE, the relatively high price of beef compared to poultry meat, and new dietary attitudes which favour lower fat consumption.

Future prospects

The ability of CEE countries to expand and develop their domestic and potential beef export markets will depend on their capacity to improve their production competitiveness, both in terms of price and quality. This can be achieved by further:
- restructuring and intensifying beef production and processing;
- modernisation and consolidation of farms;
- exploiting niche markets;
- specialisation in beef production;
- improving meat quality and product traceability;
- developing and implementing breeding policies and programmes that fully exploit the genetic potential within the region;
- promoting national trademarks and exploiting niche markets for local or special products such as the Hungarian Grey.

References

Czech Statistical Office – Český Statistický Úřad (CSÚ). Available at: http://www.czso.cz/eng/redakce.nsf/i/home

Food and Agriculture Organization Corporate Statistical Database (FAOSTAT). Available at: http://faostat.fao.org/

Statistical Office of Estonia – Eesti Statistika. Available at: http://www.stat.ee/

Statistical Office of Lithuania – Statistikos Departamentas prie Lietuvos Respublikos. Available at: http://www.stat.gov.lt/en/index/

Statistical Office of the European Communities (EUROSTAT). Available at: http://epp.eurostat.ec.europa.eu/portal/page?_pageid=1090,30070682,1090_33076576d=portal&_schema=PORTAL

Statistical Office of the Slovak Republic - Štatistický úrad Slovenskej Republiky. Available at: http://portal.statistics.sk/showdoc.do?docid=4

United States Department of Agriculture (USDA). Available at: http://www.usda.gov/wps/portal/usdahome

Dairy sector challenges and perspectives in Central and Eastern Europe

A. Rozstalnyy[1], I. Hoffmann[2] and S. Mack[2]

[1]*Food and Agriculture Organisation of the United Nations, Subregional Office for Central and Eastern Europe, Budapest, Hungary; andriy.rozstalnyy@fao.org;* [2]*Food and Agriculture Organisation of the United Nations, Rome, Italy*

Abstract

There is a dual milk production structure in most central and eastern European countries of small subsistence farms and large commercial farms. Cow numbers have declined considerably but milk output has been maintained through increased yield per cow. There have been big improvements in quality, particularly in some countries. In Poland and Czech Republic the national milk quota is a barrier to expansion. In Ukraine, Belarus, Armenia and Georgia the number of cows fell to less than half following the political changes in 1992, but in recent years, milk output has increased again. The greatest industry problem is the large number of small producers resulting in low quality milk and the declining genetic merit of the cow herd. The greatest challenge is to find the capital to modernise the industry.

Keywords: dairy cows; development; milk production

Introduction

For countries that joined the EU since 2004, changes in the dairy sector have been dramatic. The other former USSR countries can learn from the experiences of the new EU countries.

Central and Eastern EU countries

The total dairy cow population of the EU increased by approximately 4.5 million after the 2004 expansion, and by an additional 1.9 million after the accession of Bulgaria and Romania. The EU-15 had ~18.7 million dairy cows at the end of 2004, and this had increased by more than 30% by 2007 (EUROSTAT, undated). In the new EU member states, the dairy sector accounts for 7% to 20% of total agricultural output.

Herd size and production

In most of the CEE countries, the majority of the farms are very small, 1-2 cows (Figure 1 and 2). These small farms are predominantly semi-subsistence dairy farms. The smallest ones produce milk and milk derivates for their own consumption, and trade the rest as raw milk and home made dairy products like cheese, curd and yoghurt. The change in the political system in the region induced structural changes in the dairy sector. The cooperatives that used to hold large dairy herds were dissolved during privatisation. There are only some countries where the cooperative form has survived. The existence of the small semi-subsistence farms and large farms, with herds of more than 100 dairy cows result in dual structure of dairy farm size in CEE countries.

This fragmented farm size structure leads to high milk collection and other processing costs (sometimes poor quality of raw milk) on small farms, which results in low competitiveness on the EU dairy market. After EU accession, a fusion of some very small dairy farms was observed. At

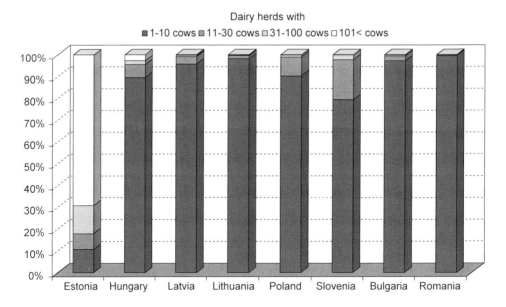

Figure 1. Dairy herd structure in some Central and Eastern EU countries.

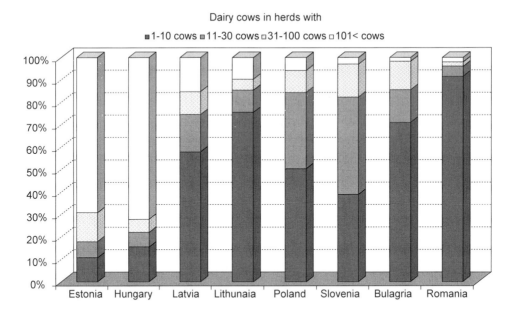

Figure 2. Dairy farm structure in some Central and Eastern EU countries.

the same time, a significant number of farmers, which owned small farms either slaughtered their herds and quit the dairy business, or switched to other livestock enterprises such as goat or sheep production. The number of herds and number of dairy cattle has declined over the past decade in most of the CEE countries (Table 1, Figure 3) but milk production did not follow this trend. Despite the decline in the number of dairy cattle milk production remained quite stable (Table 2).

Due to modernisation, concentration, EU harmonisation, and good breeding technologies milk yield per cow has increased significantly (Table 3).

Quality standards of milk have improved considerably in successful attempts to adjust to the EU standards. An example of the improvement of the quality of milk in Slovenia is illustrated in Figure 4. Similar changes can be observed in the other countries in the region.

Table 1. Number of dairy cows (×1000 head) in Central and Eastern EU countries (1998-2007).

	1997	1998	1999	2000	2001	2002	2003	2004	2005	2006	2007
Bulgaria	387.1	421.4	431.0	362.6	358.6	358.2	361.8	368.7	347.8	350.1	335.9
Czech Republic	598.0	583.0	548.0	529.0	496.0	464.0	449.0	429.3	437.1	417.3	407.4
Estonia	167.7	158.6	138.4	131.0	128.6	115.6	116.8	116.5	113.1	108.9	104.1
Hungary	379.0	384.0	376.0	355.0	345.0	338.0	310.0	304.0	285.0	268.0	266.0
Latvia	262.8	242.1	205.6	204.5	209.1	204.6	186.3	186.2	185.2	182.4	180.4
Lithuania	582.8	537.7	494.3	438.4	441.8	443.3	448.1	433.9	416.5	399.0	404.5
Poland		3,360.8	3,215.1	2,982.4	2,929.6	2,934.6	2,816.1	2,730.4	2,754.8	2,637.0	2,677.3
Romania					1619.5	1627.4		1566.4	1625.4	1639.4	1572.9
Slovakia	300.0	265.0	251.0	242.5	230.4	230.2	214.5	201.7	198.6	185.0	180.2
Slovenia	147.6	146.5	149.1	140.2	135.8	140.0	130.7	134.0	120.3	112.5	116.4

Sources: Agripolicy.net, Central Statistical Bureau of Latvia, Central Statistical Office of Poland, Czech Statistical Office, EUROSTAT, Statistical Office of the Republic of Slovenia, Statistical Office of the Slovak Republic, and Statistical Office of Estonia.

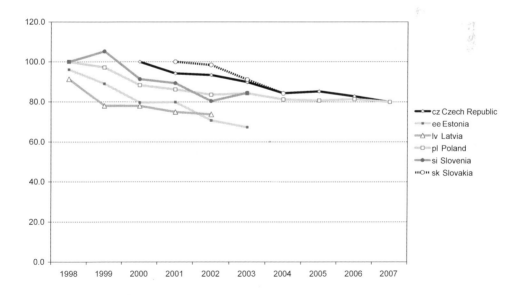

Figure 3. Number of the dairy cows in Central and Eastern EU countries (dairy cows in 1997 = 100%).

Table 2. Milk production (tonnes) in Central and Eastern EU member countries (2004-2007).

	2004	2005	2006	2007
Bulgaria	797.50	803.10	839.40	745.50
Czech Republic	2,563.22	2,543.20	2,392.50	2,445.52
Estonia	536.10	571.20	605.90	593.40
Hungary	1,541.71	1,594.00	1,448.35	1,447.73
Latvia	478.10	501.70	592.32	630.70
Lithuania	1,139.64	1,200.49	1,296.15	1,347.13
Poland	8,151.40	8,825.19	8,825.99	8,744.39
Slovakia	937.16	967.94	961.58	964.22
Slovenia	503.34	508.34	511.02	530.37

Sources: Agripolicy.net, Central Statistical Bureau of Latvia, Central Statistical Office of Poland, Czech Statistical Office, Hungarian Central Statistical Office, National Statistical Institute of Bulgaria, EUROSTAT, Statistical Office of the Republic of Slovenia, Statistical Office of the Slovak Republic, Statistical Office of Estonia, Statistical Office of Lithuania, and USDA.

Table 3. Cow yield (kg/year) in Central and Eastern EU member countries.

	2000	2001	2002	2003	2004	2005	2006
Bulgaria	3,273	2,854	3,641	3,653	3,716	3,490	3,522
Czech Republic	5,411	5,755	5,501	5,701	6,136	6,521	6,415
Estonia	4,549	5,215	4,751	5,285	5,581	5,750	5,920
Hungary	5,699	6,034	6,173	6,025	6,111	6,344	6,026
Latvia	4,002	4,136	3,880	3,827	4,208	4,332	4,361
Lithuania	3,465	3,919	3,994	4,035	4,109	4,271	4,343
Poland	3,944	4,041	4,163	4,135	4,268	4,332	4,354
Romania	2,542	2,634	2,753	2,863	3,524	3,583	3,583
Slovakia	4,335	4,793	5,199	4,826	4,892	5,314	5,275
Slovenia	4,490	4,667	5,201	4,589	4,980	4,880	4,880

Sources: Agripolicy.net, Central Statistical Bureau of Latvia, Central Statistical Office of Poland, Czech Statistical Office, Hungarian Central Statistical Office, National Statistical Institute of Bulgaria, National Institute of Statistics Romania, EUROSTAT, Statistical Office of the Republic of Slovenia, Statistical Office of the Slovak Republic, Statistical Office of Estonia, Statistical Office of Lithuania, and USDA.

Milk quota

The quota system allocates two 'national reference quantities' to each Member State, one for deliveries of milk to dairies, and another for direct sales of milk and dairy products. If either quota is exceeded, a levy is payable to the EU budget. The level of milk quota is a problematic issue in many countries. In Poland and in Czech Republic the national milk quota is a serious barrier to the further development of the dairy sector. While the total output of the Polish dairy industry is approximately 12.0 million tonnes, the output allowed under the national milk quota is only 9.2-9.4 million tonnes, which means that only about 75% of the output is commercially utilised. The EU has agreed to retain the milk quota system until 2015, after the end of the current Common Agricultural Policy reform.

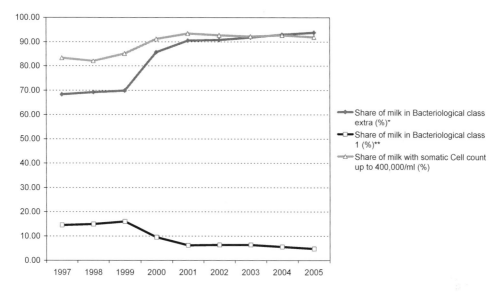

Figure 4. Share (%) of milk of different bacteriological classes in Slovenia.
Sources: Statistical Office of the Republic of Slovenia and GLiPHa.

Consumption

The consumption of milk and milk products in CEE is relatively low compared to EU-15. For example, consumption of milk and milk equivalent per capita for the following countries is Poland 250 kg, Hungary 135-155 kg, Slovakia 155 kg, Slovenia 240 kg, and Czech Republic 230 kg.

Dairy sector in Armenia, Belarus, Georgia and Ukraine

In the former USSR countries, the dairy sector accounts for up to 25% of total agricultural production. Animal production in these countries suffered seriously during the transition period in the 1990's (Figure 5). For instance, in Ukraine milking livestock have decreased to less than half in 15 years, and by the end of 2005 amounted to only 45% of the 1990 number. After that rapid decrease in the cattle population and the corresponding decrease in milk production, a gradual improvement in the situation has since been observed (Table 4, Table 5,).

The most important problems and constraints in dairy sector development include the prevalence of small-scale farms in the total raw milk supply often resulting in the production of low quality raw milk, constraints to accessing credit, low prices for milk, lack of investment in dairy farming, and underdeveloped logistics and infrastructure such as milk collection, storing and distribution. The production of feeds and fodders have decreased significantly and pastures are not well managed. The extent of artificial insemination use has sharply declined as centralised breeding farms have been abandoned and the core breeding stock have been distributed to private individuals, who are often not experienced in livestock breeding. This has led to deterioration of the genetic characteristics of cattle. In addition, high prevalence of zoonotic and transboundary animal disease such as brucellosis, tuberculosis, foot and mouth disease hinder the development of the dairy sector in some countries. Milk yield per cow is shown for 2002 to 2006 in Table 6, and from 1992 to 2007, in Figure 6.

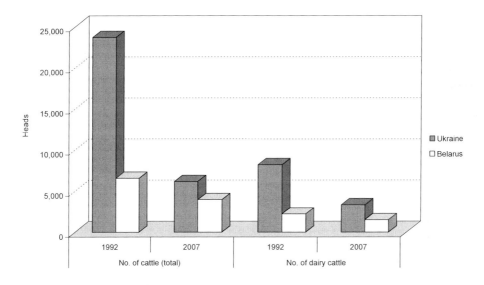

Figure 5. Number of cattle and dairy cows in Belarus and Ukraine in 1992 and 2007.

Table 4. Number of cattle and dairy cows in Armenia, Belarus, Georgia and Ukraine.

		2002	2003	2004	2005	2006
Armenia	total cattle	514,244	535,784	565,800	573,260	592,067
	dairy cows	270,107	277,000	300,000	290,069	297,060
Belarus	total cattle	4,084,500	4,005,100	3,924,000		3,989,000
	dairy cows	1,749,000	1,664,000	1,830,000		1,506,000
Georgia	total cattle	1,180,200	1,215,895	1,242,500	1,252,073	1,265,097
	dairy cows	678,270	691,500	720,000	740,752	725,349
Ukraine	total cattle	9,423,700	9,108,400	7,712,100	6,967,000	
	dairy cows	4,820,400	4,620,600	4,202,900	3,863,000	

Sources: Ministry of Agricultural Policy of Ukraine, Ministry of Agriculture and Food of Belarus, FAOSTAT, and USDA.

Table 5. Milk production (tonne/year) in Armenia, Belarus, Georgia and Ukraine.

	2002	2003	2004	2005	2006
Armenia	475,113	498,100	535,831	557,300	570,000
Belarus	4,772,500	4,682,600	5,124,100	5,650,100	5,869,900
Georgia	720,703	743,270	754,992	760,786	690,000
Ukraine	13,846,700	13,350,640	13,390,109	13,423,753	12,988,000

Sources: Ministry of Agricultural Policy of Ukraine, Ministry of Agriculture and Food of Belarus, FAOSTAT, and USDA.

Governmental and financial support, including low interest rate loans and timely subsidies, to foster investment in the development of sustainable, environmentally friendly dairy production that complies with EU/WTO quality and hygiene standards, as wells production of organic dairy products, remain the most important challenges for the dairy cattle sector of the CEE countries.

Table 6. Milk yield per cow (kg/year) in Armenia, Belarus, Georgia and Ukraine.

	2002	2003	2004	2005	2006
Armenia	1,758	1,778	1,841	1,921	1,965
Belarus	2,728	2,729	3,091	3,503	3,639
Georgia	1,062	1,054	1,037	1,034	937
Ukraine	2,872	2,889	3,125	3,419	3,308

Sources: Ministry of Agricultural Policy of Ukraine, Ministry of Agriculture and Food of Belarus, FAOSTAT, and USDA.

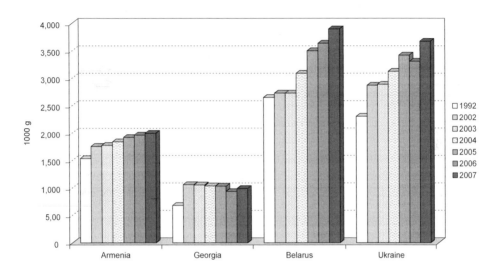

Figure 6. Milk yield per animal (kg/year) in Armenia, Belarus, Georgia and Ukraine.

References

AgriPolicy.net. Structure and competitiveness of the milk and dairy supply chains. Available at: http://www. europartnersearch.net/agri-policy/index.php?page=marketanalysis#milk

Central Statistical Bureau of Latvia – Centrālā Statistikas Pārvalde. Available at: http://www.csb.gov.lv/

Central Statistical Office of Poland - Główny Urzędu Statystyczny (GUS).Available at: http://www.stat.gov.pl/gus/ index_ENG_HTML.htm

Czech Statistical Office – Český Statistický Úřad (CSÚ).Available at: http://www.czso.cz/eng/redakce.nsf/i/home

Food and Agriculture Organization Corporate Statistical Database (FAOSTAT). Available at: http://faostat.fao.org/

Global Livestock Production and Health Atlas (GLiPHA). Available at: http://www.fao.org/ag/aga/glipha/index.jsp

Hungarian Central Statistical Office – Központi Statisztikai Hivatal (KSH). Available at: http://portal.ksh.hu/portal/ page?_pageid=38,119919&_dad=portal&_schema=PORTAL

Ministry of Agricultural Policy of Ukraine - Міністерсво Аграрної Політики України. Available at: http://www. minagro.gov.ua/

Ministry of Agriculture and Food of Belarus - Министерстве сельского хозяйства и продовольствия Республики Беларусь. Available at: http://mshp.minsk.by/structure/branches/livestock

National Institute of Statistics Romania - Institutul National de Statistica Romania. Available at: http://www.insse.ro/ cms/rw/pages/index.en.do

National Statistical Institute of Bulgaria - НАЦИОНАЛЕН СТАТИСТИЧЕСКИ ИНСТИТУТ РЕПУБЛИКА БЪЛГАРИЯ. Available at: http://www.nsi.bg/Index_e.htm

Statistical Office of the European Communities (EUROSTAT). Available at: http://epp.eurostat.ec.europa.eu/portal/page?_pageid=1090,30070682,1090_33076576&_dad=portal&_schema=PORTAL

Statistical Office of the Republic of Slovenia – Statistični Urad Republike Slovenije. Available at: http://www.stat.si/eng/index.asp

Statistical Office of the Slovak Republic - Štatistický úrad Slovenskej Republiky. Available at: http://portal.statistics.sk/showdoc.do?docid=4

Statistical Office of Estonia – Eesti Statistika. Available at: http://www.stat.ee/

Statistical Office of Lithuania – Statistikos Departamentas prie Lietuvos Respublikos. Available at: http://www.stat.gov.lt/en/index/

United States Department of Agriculture (USDA). Available at: http://www.usda.gov/wps/portal/usdahome

Developments in cattle product markets and product prices

K.J. Peters

Humboldt-Universität Berlin, Institute of Animal Sciences, Dept. of Animal Breeding in the Tropics and Subtropics, Haus 9, Philippstr. 13, 10115 Berlin, Germany; k.peters@agrar.hu-berlin.de

Abstract

World cattle stocks continue to shift towards Asia and Latin America with major gains in China, Brazil, Central Asian countries and Oceania. Largest stock reductions occurred in Russia, Eastern European countries, but also in EU and North-America. World beef production rose by 2.3% in 2007 and is expected to rise in 2008 to 68 million tonnes with 56% produced in developing countries. All regions where beef is produced from the dairy herd show large reductions in output (Eastern European countries and Russia -35.8% and -33,3%, EU -8,4%) while in North-America (Canada minus, USA plus) beef benefits from favourable exchange rates. Beef production in Latin-America continues to grow while in Argentina the recently introduced export taxes reduces the competitiveness of beef production beyond the domestic demand. In Australia, the output of beef will depend on the grain prices and past increases are less likely to be sustained. International beef trade amounts to about 7.1 million tonnes in 2007, and the market moved out of drought periods in Oceania and BSE incidences curtailing trade. Major importing countries besides the EU are Russia, USA/Canada and China; major exporting countries are EU, EEC, Oceania, and India. Largest increases in imports occurred in EEC and Russia, but only in Russia are beef imports expected to grow due to demand increases and unfavourable domestic production conditions. Beef exports from EEC are linked to trade policies in neighbouring countries, mainly Russia, where currently major trade contracts are established with Brazil and USA. International beef prices increased by almost 7% in 2008, due to growth in global demand and limited export supplies from traditional producers and from beef from the dairy herd. Structural factors, price policies and production inefficiencies are the major reasons for the down scaling of cattle and beef production in Russia and EEC. Total milk production reached 676 million tonnes in 2007 and is expected to grow by 2.5% in 2008. Largest expansions occurred in developing countries, putting their share of global production at around 47%, but in the major exporting countries, responsible for 80% of global export, milk production increased by only 1% following a decline of 0.7% in 2007. Strong expansion in Belarus, Argentina and the US, and marginal growth in the EU and Ukraine, was partially offset by declines in Oceania. The price increase for most dairy products in 2007, due to increasing demand in Asia and Russia, effective policies in EU to reduce overproduction, and depletion of most stocks, is unlikely to be sustained, since previous importers have expanded their own production. How world milk prices affect local production and competitiveness very much depends on the impact of regulation policies active in most dairy countries around the globe. Observed costs and prices vary by a factor of 2.5 and still do not have a major impact on the competiveness of dairy production. Other factors are linked to structural sector development, production and processing efficiency. Eastern European countries and Russia depend to a large extend on supporting policies to allow the modernisation of the physical and institutional dairy value chain, to stimulate investment in efficient production units and marketing infrastructure, and to enhanced human skills for quality production.

Keywords: cattle population, beef, dairy, production, trade, prices

Introduction

Global meat production in 2007 reached almost 275 million tonnes (million tonnes) of which 24.5 % derives from bovines (cattle and buffaloes). With an annual growth in beef production of around 1% *per annum* (p.a.) the relative contribution of beef to global meat production is steadily decreasing. In 2007 poultry and pig meat production contributed 32.6% and 35% respectively, with expected growth rates of 3.8% and 1.8% for 2008 (FAO, 2008a). Despite the fact that meat production with monogastric species can be expanded faster and has a better feed efficiency, beef production based on conversion of roughage feeds with some additional concentrate finishing, continues to have a place in global meat production in order to meet consumer demands.

According to FAO data global milk production is estimated to have expanded by 1.8% to 676 million tonnes in 2007, and growth in 2008 is expected to reach 2.5 %, as producers responded to high prices in 2007. The largest expansion of production occurred in developing countries lifting its global share in milk production to 47.5% (FAO, 2008b).

Only 8% to 9% of the global milk volume is traded of which 80% is covered by six leading exporting countries. Changes in production conditions in these countries and short term changes in demand for dairy products can lead to major shifts in the global market situation.

This paper deals with trends and developments in cattle stocks, the dynamics in beef production and beef trade, and trends in dairy production and markets.

World cattle stocks and changes

World cattle stocks during the last decade rose modestly by 1.32 million to 1.38 million or about 0.5% p.a. Most cattle are kept in Latin America, China and India and stock changes show gains and losses. Stocks declined most drastically in Russia and the Eastern European transition countries. A modest decline is also noticeable in high income countries (EU, USA/Canada), while major stock expansions occurred in countries of the southern hemisphere, and mainly in countries with fast economic growth such as Brazil (+3.0% p.a.) and China (+1.8% p.a.). Central Asian countries moved towards a stabilised cattle sector with an increase (+1.1% p.a.) over the last 10 years. Australia continues a phase of restocking after a prolonged drought and final stocks will depend on the development of beef markets but also the development of feed prices (Table 1).

Table 1. Cattle stocks and cattle meat production (FAOSTAT, 2008).

Country	Cattle stock (million)		Meat production in million tonnes	
	2006	Changes since 1996 (%)	2006	Changes since 1996 (%)
Argentina	50.8	-0.1	3.0	+10.6
Brazil	207.2	+30.9	6.1	+25.7
Australia / New Zealand	38.2	+8.0	2.8	+16.8
China	117.8	+18.4	7.2	+115.2
India	180.8	-9.9	1.3	-2.6
USA / Canada	111.5	-6.6	13.3	+4.2
EU 27	90.4	-9.1	8.0	-8.4
Russia	21.4	-45.9	1.8	-33.3
East European countries (Ukraine, Moldova, Belarus)	10.8	-53.7	0.9	-35.6
Central Asian countries	17.1	+11.4	1.1	+13.8

A new policy direction in Russia is now strongly encouraging the import of live dairy and beef cattle for restocking and revitalising the cattle sector (in 2007 141.2 million head). Most of the live importations (dual purpose cattle, dairy cattle) have traditionally been supplied by EU countries but, small numbers were derived also from Australia. Live beef cattle were obtained from Canada and in future also from the USA (Hansen *et al.*, 2008; USDA, 2008a,b).

Trends in the beef sector

Despite the fact that beef production contributes a steadily smaller share to the global meat supply it is growing proportionally faster than meat consumption in developed countries, indicating a special preference for beef with rising purchasing power and with a high level of overall meet consumption (Table 2).

Production

After a decline of 1.3% in 2006, world bovine meat production rose again by 2.3% in 2007, mainly due to lifts of import bans or restrictions (Japan, Korea, Russia). Leading beef producing countries are USA/Canada, EU, China, Brazil, and Australia/New Zealand. Expansion of beef production over the last decade (+1.15% p.a.) have been below human population increases and is expected to grow by 2.7% ending at 7.6 million tonnes in 2017 (FAPRI, 2008). During the decade from 1996 to 2006 major expansions occurred in China, Brazil, Australia/New Zealand, Central Asian Countries, and Argentine countries (Table 2).

Beef meat production in North America is expected to remain virtually unchanged. Expansion in the United States will be offset by a 6% decline in Canada due to impacts of the implementation of the Country of Origin Labelling (COOL) regulation by the United States. The increase in USA beef output which is partly due to its depreciating currency, has increased its competitiveness. Moreover, high supplies of distiller-dried grains from the production of ethanol have helped to lessen the impact of higher feed costs (FAO, 2008a)

In Brazil, the largest producer in Latin America, the 5% growth in 2007 is expected to be reduced to 2.5% in 2008, if no alternative markets are found to offset the effect of the new restrictions imposed

Table 2. World meat markets summary (FAOSTAT, 2008).

World balance	2006	2007[a]	2008[b]	Change: 2008 over 2007
	million tonnes			
Production	271.5	274.7	280.9	2.3
Bovine meat	65.7	67.2	68.0	1.1
Supply and demand indicators				
Per capita meat consumption:				
World (kg/year)	41.6	41.6	42.1	1.1
Developed (kg/year)	81.1	82.4	82.9	0.7
Developing (kg/year)	30.7	30.5	31.1	1.8
				Change: Jan-Apr 2007:2008
FAO Meat price index				%
(1998-2000 = 100%)	115	121	131[c]	10

[a] Estimated.
[b] Forecast.
[c] Jan-Apr 2008.

by the European Union on imports from that country due to product safety concerns related to animal diseases. Since Russia has shown a major interest in contracted imports of beef from Brazil, current trends are still set for a comfortable growth (Hansen *et al.*, 2008; USDA, 2008a). Argentina, with its well established export oriented beef production system, is already suffering from the imposition of government export taxes, strongly reducing the economic competitiveness of producing beef for export, and shifting the whole agricultural sector to domestic markets. Other major beef producing countries in South America (Chile, Columbia, Paraguay and Venezuela) are expected to expand their production by around 5% (FAO, 2008a).

Steady expansion of cattle stocks, improved management practices, and strong government support are the basis for the continued expected growth of beef output in China (+3% in 2008).

Beef output in Australia is expected to decline by 3.3% in 2008, due to major efforts to rebuild cattle stocks, and due to changes in competitiveness of grain feeding in feedlots due to higher feed costs in 2007. Favourable product prices for beef will boost beef production in New Zealand, although further development is linked to the relative competitiveness of dairy and beef production.

Beef production in the EU continues its negative trend, as the number of dairy cattle is declining due to major milk yield increases, and as the number of beef cattle remains stable. The huge reduction in beef production from 1996 to 2006 in Eastern European countries (-35.6%) and Russia (-33.3%) is linked to major transformation processes in the livestock sector in these countries. While in Belarus and Ukraine the restructuring process has gained momentum and revitalised beef output, in Russia beef production continues to decline, due to structural problems in the agricultural sector and reduced competitiveness of beef in the domestic meat market (Hansen *et al.*, 2008; USDA, 2008a).

The dynamic and interacting trends in beef production in different regions and countries clearly suggest that beef output from high yielding dairy herds is declining, while dual purpose dairy herds with lower milk yields are still a major source of beef, and beef from specialised beef suckler-cow systems are showing the largest expansions. The increased specialisation and intensification of dairy production, thus, strongly affects beef output and the trends in beef trade. Countries with larger farms (high ratio of land to humans) and favourable ecological conditions for beef cattle production will be the future beef producing countries (Table 3).

Table 3. Origin of beef (Deblitz *et al.*, 2004).

From dairy herd to the suckler cows!
- 'Dairy countries' (<25% of cows are suckler cows)
  - Poland, Pakistan, Hungary, Czech Republic and Germany
  - Decreasing beef systems
- 'Mixed countries' (25 to 75% of cows are suckler cows)
  - New Zealand, Austria, France, Ireland and Spain
  - Balanced beef systems
- 'Beef countries' (>75% of cows are suckler cows)
  - USA, Canada, Brazil, Australia, Argentina and Uruguay
  - Expanding beef systems

Beef trade

International beef trade amounted to about 7.1 million tonnes in 2007 and is forecast at 7.2 million tonnes in 2008. This represents about 9.5% of total global beef production. The world beef market has been affected by a series of droughts in Australia and by the BSE incidents in North America that resulted in the imposition of bans by many importers. As these are being progressively lifted, trade in beef is resuming a more normal pattern.

According to FAOSTAT (2008), Eastern European countries have gained first position as beef exporters, followed by Australia/New Zealand and India. The largest import/export beef market is still the EU. Most of this trade is within the EU, and the import of beef from Brazil has declined substantially due to EU demands on traceability and production standards. Other South American countries are not in a state to fill the gap since EU standards cannot be met, or as in the case of Argentina, due to loss of competitiveness (Table 4).

Major downward shifts in exports during recent years are noticeable for USA/Canada to mainly Japan and South Korea as a result of BSE incidences. The strong Euro, high internal prices and decreased imports from Brazil, however, will discourage exports from the European Union. Canada's beef shipments are also expected to fall, negatively affected by the introduction of the Country of Origin Labelling legislation in the United States, while exports from the USA are anticipated to rise, sustained by a weak dollar and the progressive lifting of import bans by its traditional importing partners.

Russia is the largest net importer of beef over the last few years (Table 4). A 12 year decline in domestic beef production because of decreased cattle and feedstock, increased grain prices, and generally negative profitability of cattle and beef production has discouraged new investors. In 2006, the imports of frozen beef totalled 647,200 tonnes, and increased by 25% in 2007. Supplies to the Russian beef market shifted from European sources (disease related import bans) to mainly South American countries (Table 5). Recently, imports started again from Ukraine and also from Poland (USDA, 2008b).

The other large net importer of beef is China, also after a long period of exporting beef, though at a rather low levels. With continuous strong economic growth and raising domestic beef consumption it is assumed that imports will steadily increase in the years ahead. Despite remarkable progress in expanding cattle stocks and productivity, beef production is limited due to limited good pastureland and high opportunity cost for feed (FAPRI, 2008).

The strong Euro, high internal prices and decreased imports from Brazil, however, will discourage exports from the EU, while exports from the USA are anticipated to rise, sustained by a weak dollar and the progressive lifting of import bans by its traditional importing partners (FAO, 2008a).

Exports from Brazil reflect domestic production growth and the opening of non-traditional markets, such as Russia, to offset import restrictions imposed by the European Union, while exports from Argentina have drastically declined and are expected to stabilise at a rather low level, due to the

Table 4. Cattle meat trade 1995-2005 (×1000 tonnes) (modified from FAOSTAT, 2008).

Country	1995		2005		Δ (%)	
	Import	Export	Import	Export	Import	Export
Argentina	3.02	42.71	2.57	9.24	-15	-78
Brazil	60.33	0.01	3.05	1.66	-95	+165
Australia / New Zealand	0.30	86.09	0.24	88.69	-20	+3
China	8.77	1.79	11.05	1.52	+26	-15
India	-	27.87	-	31.25	-	+12
USA / Canada	114.73	177.92	23.35	26.66	-80	-85
EU 27	1,019.58	1,116.49	842.87	893.59	-17	-20
Russia	149.70	0.17	259.39	0.00	+73	-100
East European countries (Ukraine, Moldova, Belarus)	1.95	42.68	5.85	95.09	+200	+123
Central Asian countries	8.45	0.85	0.09	0.08	-99	-91

Table 5. Sources of beef imports to Russia (World Trade Atlas, 2008).

Country	2007 (%)	Δ 2005-2007 (%)
Brazil	65.8	+168.9
Argentina	15.7	-40.0
Paraguay	8.5	+27.3
Ukraine	5.1	-41.5
Uruguay	1.6	+266.7
Germany	1.0	-59.3
Ireland	0.6	-82.3
Others	1.6	-69.6
Total	100.0	+10.4

taxes levied on exports; exports by other Latin American countries, Chile, Paraguay, Uruguay will fill the gap.

Buffalo meat exports from India have steadily grown in the past decade and are likely to rise in 2008, in response to strong import demands from Indonesia, Malaysia, the Philippines and countries in the Near East (FAO, 2008a).

Beef prices

Driven by high feed cost and rising global import demand and temporal export limitations by major exporting countries, the FAO bovine price index rose by around 7% from 2007 to 2008 (FAO, 2008a). The general price trend over the last 5 years shows a steady increase for beef and also demonstrates that beef compared to monogastric meat is expensive, but is surpassed by lamb meat (Figure 1).

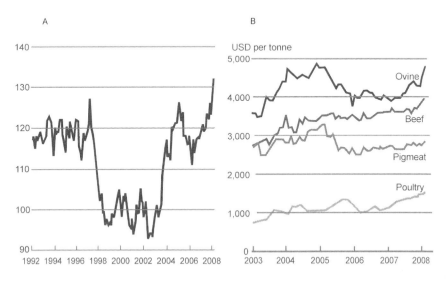

Figure 1. (A) FAO international price index for meat products (1998-2000 = 100%). (B) FAO international price index for selected meat products (FAO, 2008a).

National price levels of beef strongly reflect respective production costs. Differences between export oriented countries with low production costs and those depending on imports, and with only a minor domestic production, amount to 200% (Figure 2).

The pressure on market prices deriving from differences in production costs generally leads to respective policy measures to maintain beef production also in regions with a high production cost. As the IFCN-Beef has demonstrated, 4 fold cost differences between farms in response to farming conditions, land endowment and land value, and intensification needs exist (Deblitz *et al.*, 2004).

Given the impact of ecological factors on beef output, the global supply level, the range of short or medium term policy decisions related to beef production conditions, beef export and import regulations, domestic trade and price policies, the growing consumer demands related to SPS and traceability, the changing consumption pattern for beef in relation to major disease incidences, the world market price of beef will always show strong variations, especially since the volume of globally traded beef is only a small fraction of the total production.

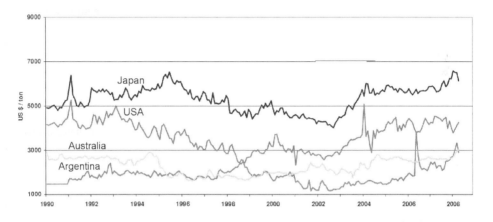

USA - Frozen beef, export unit value.
Argentina - Export unit value of chilled and frozen beef cuts.
Japan - Import price for chilled beef cuts.
Australia - Up to Oct02 : cow forequarters frozen boneless, 85% chemical lean, c.i.f. US port (East Cost) ex-dock;
From Nov02 : chucks and cow forequarters.

Figure 2. Bovine meat: international prices (FAO, 2008b).

Trends in the dairy sector

Dairy production

Global milk production reached 676 million tonnes in 2007, up by 1.8% over 2006. Production growth in 2008 is expected to be around 2.5% as producers respond to better prices in 2007. The leading dairy production regions are EU (22.7%), India (14.3%), USA/Canada (13.4%), China (5.4%), Russia (4.6%), Brazil (3.8%), Oceania (3.6%), Eastern Europe (2.9%), and Central Asian countries (1.9%). Major expansions of dairying during the last decade (Table 6) occurred in countries with either fast growing economies and low per capita dairy consumption levels, or large increases in demand for dairy products, such as China (+258%), Central Asian countries (+44%), India (40.4%), and Brazil (32.4%). Among the leading dairy countries with high levels of consumption of dairy products only

Table 6. Cattle milk production level and development (million tonnes) (FAOSTAT, 2008).

Country	2006	Changes since 1996 (%)
Argentina	8.1	-11.4
Brazil	25.4	+32.4
Australia / New Zealand	24.7	+30.3
China	36.4	+257.8
India	95.7	+40.0
USA / Canada	90.6	+18.5
EU 27	153.8	+1.7
Russia	31.3	-12.5
East European countries	19.8	-35.6
(Ukraine, Moldova, Belarus)		
Central Asian countries	12.7	+44.0

Oceania (+30.3%) and North America (+18.5%) dairy production had a sizable expansion, while in the EU the quota system kept dairy output almost constant (FAO, 2008a).

Production trends during the last few years highlight the continuous expansion in so called developing countries reaching a global share of almost 48% in 2008. The dairy sector in China expanded by 8% and 9% during 2006 and 2007, and in India and Pakistan by 3% and 4%, respectively. The strong impact of purchasing power on demand for dairy products in countries with a fast growing economy and yet low consumption levels will continue to either provide incentives for expanding the domestic dairy sector or for increasing the rate of imports of dairy products. As shown in Figure 3, there appears a dependency between per capita consumption of milk and growth rates of dairy production. This stipulates continuous high growth rates in all countries with emerging economies and shifts of dietary habits. Change of purchasing power and lifestyle is well connected with urbanisation and over-proportional economic growth in urban areas, as the data from China convincingly demonstrate. Dairy output in the traditionally export oriented dairy regions of the world (Europe, Oceania, Brazil/Argentina) will not change very much, and growth in some regions/countries is offset by declines in others, which does affect export trade conditions. The strong growth during 2008 in dairy production in Belarus (+3.9%), Argentina (+6.0%), the USA (+2.7%), and the marginal expansion in the EU (+0.6%) and Ukraine (+0.3%) is expected to be counterbalanced by a major decline in Australia (-3.5%) and New Zealand (-4.5%) (FAO, 2008a).

In Eastern European countries, the dairy sector over the last five years remains stagnant in Caucasian countries, shows a continuous revitalisation in Central Asian countries, appears to be expanding rather fast in Belarus (+4.0% during 2007), turns to recovery in Russia with a recent growth of +2.0% p.a., and exhibits a very mixed situation in Ukraine, where after considerable stabilisation, the export of dairy products to Russia were impacted by quality problems (-7.0% in 2007) (Figure 4).

Dairy trade

Only 8% to 9% of the global annual dairy production is traded in international markets. In 2007 the volume of internationally traded products dropped to 5.6% milk equivalent. The most important dairy products traded are butter, cheese, milk powders, casein and condensed milk. The EU and New Zealand remain the major dairy exporters accounting for over 30% each of all exports, followed by Australia. Conditions in these countries will largely affect the global trade (Figure 5).

The EU is further reducing its export ability as an effect of the quota system and in Oceania dairy output is reduced by recurrent drought (mainly in Australia). Other traditional or emerging dairy

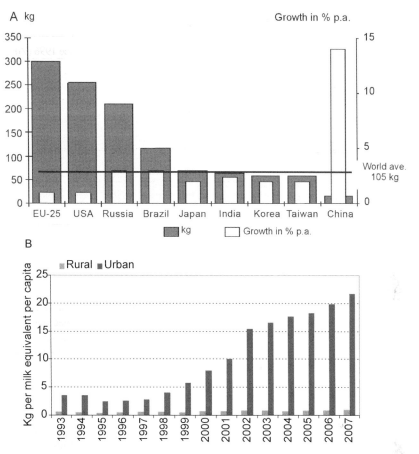

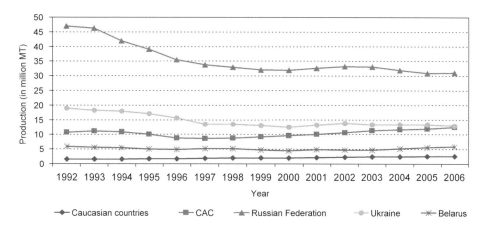

Figure 3. (A). Per capita consumption of dairy products (Danish Dairy Board, 2008c). (B). Urban and rural dairy products consumption rates expressed in milk equivalent (ME) in China (Danish Dairy Board, 2008c).

Figure 4. Milk production in Eastern European countries (FAOSTAT, 2008).

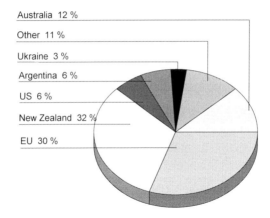

Figure 5. Exporters' share of world trade in 2006 (milk equivalent) (Dairy Australia, 2008).

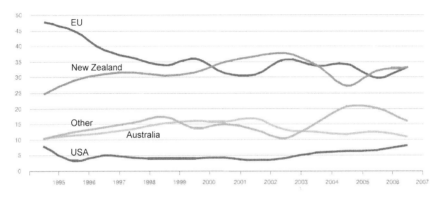

Figure 6. Share of export trade in milk equivalent (%) (Dairy Australia, 2008).

exporters are not fully able to offset this effect due to comparable low supplies to the world market (Figure 6).

Major importing countries were for many years the emerging dairy consumer nations in North Africa, Near East, South East Asia, East Asia and China. In the light of the reduced volume of traded dairy products linked with dramatic price increases many importing countries have reduced imports by around 5%, raised domestic dairy prices, and created incentives for domestic production. Import volumes contracted by developed countries have remained robust (FAO, 2008a).

The product structure of traded dairy commodities is changing. Export of a traditional product such as butter has declined by 26% during 2004 to 2008 in general, and in the EU from 355,000 tonnes in 2004 to 105,000 tonnes (-29.6%) in 2008. Only Belarus is expanding its butter export. The Russian Federation appears as the dominant importer of butter and shows expanding trends as long as the domestic dairy sector is unable to match the growing consumer demand in the country. Other traditional importers are North Africa, South East Asia and the Middle East Gulf States.

Skim milk powder (SMP) exports from the EU, Australia and New Zealand have also declined by almost 33%, which will give the USA first place in exporting SMP. Major importers are countries in South East Asia, North Africa, and Central and Southern America.

The global trade in whole milk powder (WMP) is constrained by insufficient supplies despite strong import demands for the use of reconstituted milk and other products. Main exporters are New Zealand and Australia and until recent the EU, main importers are Middle East and countries in Central and South America.

The international trade in cheese is very robust and market demand appears to be growing (+3.6% in 2007) despite considerable price increases. The EU is the world leader in the global cheese market with 35% of the total market share, and the amount of milk being used for cheese production is increasing. Australia and New Zealand, the second largest cheese exporters (32% of international traded cheese in 2007), are expected to reduce their exports due to a shortage in milk supplies. The USA is a major cheese importer under preferential trade agreements with the EU, New Zealand and Australia, but has recently increased its exports. Other importers are Japan, Middle East and North Africa.

Dairy prices

Rising purchasing power, population growth and limited agricultural production resources will sustain higher food prices and especially those of livestock origin. Dairy production should benefit from this trend since efficient production is much dependent on favourable ecological conditions and the ability to improve management and to connect to markets. The rising demand for dairy products in emerging economies with climates less favourable for dairying can hardly be met by expanded domestic production, with only moderate yields. Thus, dairy production in temperate regions with a favourable ecology remains a beneficiary of rising global demands for dairy products.

Dairy product prices rose faster than other agricultural commodities during 2006 and 2007 as a result of diminishing stocks in major exporting countries. Underlying reasons for the dramatic changes were the concomitant effects of higher feed prices, drought effects, and the effect of the EU Common Agricultural Policy diminishing surplus production through the rigid quota system (Morgan, 2008). The FAO index of dairy product prices (1998-2000 = 100) rose to 266 in April 2008, which represents a 25 % increase in one year (Figure 7).

SMP experienced the highest price rise and induced increased production which finally caused a marked adjustment in prices. But other products experienced a price reduction (WMP 8%, butter 5%, cheese 8%) during early 2008.

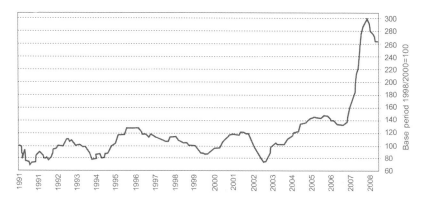

Figure 7. Monthly indices of international prices of selected dairy products (Export trade value weights for the world) (FAO, 2008b).The index is derived from a trade-weighted average of a selection of representative internationally-traded dairy products.

Although it is expected that dairy prices will remain robust, there are a number of uncertainties. Factors supporting a lower price are related to the ability of importing countries to expand their own dairy sector and cut imports, or allowing export countries access to the global market without prohibitive export taxes as in the case of Argentina.

Other uncertainties are related to changes in feed cost as caused by expanding land-use for bio-energy, impairing the competitiveness of dairy production. The propagation of bio-fuel is linked to overall energy prices but is also linked to respective policies to reduce dependency on fossil fuel. Policy changes in this area affect conditions for dairy production. The current cost level of fuel and feed are markedly affecting the competitiveness of intensive dairy production and could lead to major culling and a reduction in dairy supplies to the market, which eventually would again lead to increasing product prices in the global market.

The actual milk prices received by dairy producers show very large variation in different countries and under different production circumstances within countries. As the IFCN Dairy report of 2005 reveals (IFCN, 2005), farm gate prices differ by more than 100% with extremely low prices in countries with emerging dairy sectors and lower factor cost for land and labour and rather high prices in countries with intensive dairy farms (Figure 8).

Price increases such as those during 2007 may not automatically lead to higher farm gate prices. The Danish Dairy Board reports on the adjustments in different EU Countries and the USA which varied from 16% in Finland to 34-38% in Germany and Holland and up to 45% in UK and Ireland (Danish Dairy Board, 2008a). The market position of dairy producers in different countries is influenced not only by the market power of the ever decreasing number of dairy processing plants but also by the increasing discounters in the retail business. Thus, margins between farm gate prices and consumer prices (Figure 9) do vary considerably across countries, as the work by IFCN so clearly demonstrates (IFCN, 2005, 2006).

Dairy prices paid to the farmer and prices obtained in export markets are very difficult to rationally explain, since they are influenced not only by production related cost factors, but also by domestic /national price support arrangements, which are more often related to rural social policies than to anything else.

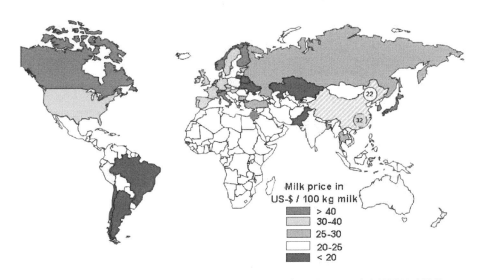

Figure 8. Farmers' milk prices (US$ per 100 kg milk - 4% fat, 3.3% protein) (IFCN, 2005).

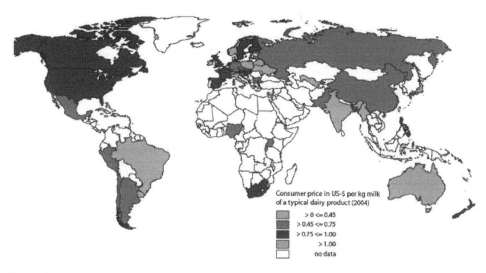

Figure 9. IFCN estimate for consumer prices for milk (US$ per kg milk in 2004) (IFCN, 2005).

In markets with quota regulations for managing imports, the realised returns from sales of dairy products are affected by the respective protected price structures in the importing countries, which are normally well above the international price level. Other distortions of international prices are caused by preferential trade agreements or general export subsidies of export oriented countries such as the EU and USA.

Exchange rate fluctuations, level of market stocks of products in the EU and USA, modifications in subsidy regulations and payments, and the volume of food aid supply can also impact on price developments. The marketable volume of dairy products is further influenced by seasonality of production and the severity of recurrent droughts, as it mainly occurs in the southern hemisphere, and thus impacts on national and international market prices.

Conclusion

Cattle are the most important large livestock species for supplying meat and milk to feed the ever growing human population. Expansion of cattle populations are occurring mainly in regions with growing product markets and/or with favourable production conditions, i.e. in emerging economies and in countries with production reserves. In traditional dairy producing countries with high yields, the number of cattle is stagnant and product market opportunities are met with yield increases.

Beef production grows at just above 1% p.a. and is tending to move to countries with production reserves in the southern hemisphere, while the highly specialised dairy systems loose their role to supply feeder cattle for the beef sector. International trade in beef flows from major export regions such as Oceania, South America, and India to deficit countries particularly Russia, China and the EU. Major changes in cattle stocks and beef production occurred during the last 10 years in transition countries of Eastern Europe and Central Asia, indicating difficulties in overcoming structural changes in the livestock sector, in institutions of the cattle value chain, and in regaining a competitive cattle sector, but also in adjusting to new quality standards and trade policies of neighbouring countries. Beef prices have not changed dramatically but show a large variation across countries.

Milk production shows a steady expansion of 1.7% p.a. with considerable differences according to the prevailing consumption level. The largest expansion is occuring in countries of the southern hemisphere and in countries with strong economic growth fuelled by growing purchasing power and economic strength of urban areas. The leading dairy producing regions are the EU, USA, India and Oceania and the leading export countries are the EU and Oceania.

Exceptional price increases for dairy products were related to the growing import demand of deficit countries, and to reduced product stocks in the EU. A general price adjustment seems correlated to energy prices and its effects on general production costs and on the emergence of bio-fuel. Dairy prices will continue to fluctuate considerably due to price disturbing effects of policies.

References

Dairy Australia, 2008. Dairy production and trade. Available at: www.dairyaustralia.com.au/content/view/197/83/.

Danish Dairy Board, 2008. Mejeriforeningen. Per capita consumption of dairy products. Available at: http://www.mejeri.dk/smcms/danishdairyboard_dk/Policies/International_dairy/Per_capita/Index.htm?ID=7925.

Deblitz, C., Charry, A.A. and Parton, K.A., 2004. Beef farming across the world: an expert assessment from an international co-operative research project (IFCN). Available at: http://www.csu.edu.au/faculty/science/saws/afbmnetwork/efsjournal/volume1/number1/EFS_Journal_v01_n01_01_ClausDeblitz_et_%20al.pdf.

FAPRI, 2008. FAPRI 2008 U.S. and World Agricultural Outlook Database. Food and Agricultural Policy Research Institute. Available at: http://www.fapri.iastate.edu/tools/outlook.aspx.

FAO, 2008a. Food outlook, global market analysis. June 2008. Food and Agriculture Organization of the United Nations. Available at: http://www.fao.org/docrep/010/ai466e/ai466e00.htm.

FAO, 2008b. Trade and markets. Food and Agriculture Organization of the United Nations. Available at: http://www.fao.org/es/ESC/en/15/162/highlight_176.htm.

FAOSTAT, 2008. Statistical service of the Food and Agriculture Organization of the United Nations. Available at: http://faostat.fao.org/.

Hansen, E., Maksimenko, M. and DuBois, C.A., 2008. Russian Federation livestock and products semi-annual report 2008. USDA Foreign Agricultural Service. Available at: http://www.fas.usda.gov/gainfiles/200803/146293870.pdf.

IFCN, 2005. Annual Report, 2005. IFCN Dairy Research Center.

IFCN, 2006. Annual Report, 2006. IFCN Dairy Research Center.

Morgan, N., 2008. Dairy prices, policies and potential opportunities for smallholder in Asia. An APHCA Brief, 2008. Available at: http::/www.fao.org/docrep/010/ai465e00.htm.

USDA, 2008a. Gain Report Number RS8014, release 0120.08. United States Department of Agriculture, Foreign Agricultural Service. Available at: http://www.porkworld.com.br/img/File/41.pdf.

USDA, 2008b. United States Department of Agriculture. Russia allows for importation of U.S. Livestock. 05. Jun. 2008. 04. Aug. 2008 http://www.usda.gov/wps/portal/!ut/p/_s.7_0_A/7_0_1OB?contentidonly=true&content id=2008/05/0120.xml.

World Trade Atlas, 2008. Importe nehmen zu. ZMP Welt Agrar Markt, Januar 2008.

Analysis of developments in new EU member states based on the dairy quota situation

A. Kuipers[1], A. Malak-Rawlikowska[2], M. Klopcic[3] and J. Sataite[4]

[1]Expertise centre for Farm Management and Knowledge Transfer (Agro Management Tools), Wageningen UR, the Netherlands; abele.kuipers@wur.nl; [2]Warsaw University of Life Sciences, Faculty of Economic Sciences, Warsaw, Poland; [3]University of Ljubljana, Biotechnical Faculty, Slovenia; [4]State Enterprise Agricultural Information and Rural Business Centre, Milk Quota Accounting Bureau, Vilnius, Lithuania

Abstract

The world dairy situation was recently (2007) quite positive in comparison to the period 2002-2006. Demand for dairy products is growing, especially because of the rapidly increasing demand in Asian countries. In 2007 and early part of 2008, a substantial increase in milk prices was seen in the EU, as well as an increase in some cost factors (feed and energy). Nevertheless, 20 out of the 25 EU countries did not reach or just reached their national milk quota in year 2006/2007. The overall milk production volume in the EU is stable and export is decreasing. On the contrary, New Zealand, South America and USA are strengthening their export position on the world market. This trend has caused a discussion in EU about the future of the quota system after 2014; keep this production restricting instrument or choose an open market situation. The situation in some EU countries will be described in detail. As case studies, the situation in countries of Eastern Europe with relatively small herds, i.e. Poland, Slovenia and Lithuania are examined. Regionalisation of milk production is compared against the quota system in use: a regional system versus a national system. The structural developments in the various regions in these countries are also compared. The decoupling of EU subsidies from production has made the choice of strategy more an economical farm management decision than before. The dairy sector in Europe is becoming a business in a risky market environment as was experienced by the pig, poultry, vegetable, and flower sectors for many years. Because of fluctuating milk prices in recent years and uncertainty about quota system, prices of quota have started to fluctuate. It will be a huge challenge for the sector to cope with this market situation in a sustainable way. Robustness of the farm becomes a factor of more importance.

Keywords: milk quota, regional and national systems, chain developments, Eastern EU countries

Introduction

In this study, first the EU dairy policy will be briefly described. The quota system has been the backbone of the EU dairy policy since 1984 to balance supply and demand. Some characteristics of the quota system will be outlined, which have a relationship to structural developments of the sector in the various countries. Then the application of the quota system since 2004 in the new EU countries will be described. Emphasis will be given to structural development aspects. This will be done more extensively for three countries, namely Poland, Slovenia and Lithuania. Poland is chosen because it is the largest dairy country that entered the EU in 2004. It has relatively small farms. Slovenia has a mountainous landscape and also has small farms. In both countries a form of regionalisation takes place in applying the quota system. Additional attention is given to Lithuania as one of the Baltic countries: this country has a nationally applied quota system, i.e. no regionalisation takes place. This yes/no regionalisation was also a reason behind the selection of this Baltic country. Also, extra information is provided about Romania and Bulgaria as the last countries that entered the EU. Thus, the study covers a wide spectrum of the existing diversity in European dairy production.

EU Agricultural policy

Detailed information on the EU dairy sector can be found in the fact sheet 'Milk and milk products in the European Union' (EC, 2007a). This fact sheet examines the dairy sector in facts and figures, explains the role of the EU's Common Agricultural Policy (CAP) in relation to milk production and marketing, and highlights the main factors that will influence its future.

According to this fact sheet, a common market organisation (CMO) for milk and milk products was set up in 1968. Although over the years, the dairy CMO has changed fundamentally, it still operates in three areas:
- internal market support;
- using trade instruments;
- making direct payments to farmers.

Internal market support

'Safety-net' intervention

Nowadays, public intervention (buying into storage) for butter and skimmed milk powder is limited. Intervention agencies may only buy in butter during the period 1 March to 31 August of any year. There is also a maximum or threshold on the quantities of butter offered for intervention. This threshold was 50,000 tonnes in 2006, 40,000 tonnes in 2007 and 30,000 tonnes for 2008 and subsequent years. If the threshold is exceeded the Commission may suspend conventional intervention buying and continue buying using a tendering procedure.

In 2003 it was agreed that the butter intervention price would be reduced by 25% over a four-year period, beginning on 1 July 2004, the four reductions being three times 7% plus a final cut of 4% in 2007. Moreover, the actual buying in price is only 90% of the intervention price (i.e. €221.75 per 100 kg on 1 July 2007).

Skimmed milk powder (SMP) intervention was only open between 1 March and end-August each year, for a maximum quantity of 109,000 tonnes. Beyond this quantity, intervention may be suspended and may be replaced by a tender procedure. The SMP intervention price was reduced by 15% over a three-year period, with reductions of 5% in each of 2004, 2005 and 2006, resulting in the following price levels: €205.52/100 kg in 2003/2004, reducing to €174.69/100 kg from 1 July, 2006

Disposal of dairy products on the internal EU market

In order that a healthy market balance is maintained, the EU dairy industry continues to have access to measures to ensure the competitiveness of their dairy products on the internal market. Various schemes for dairy products on the EU market still play a role in the dairy regime, though spending has been reducing in recent years in most cases. The main subsidised disposal schemes are:
- cream, butter and concentrated butter for non-profit organisations, for commercial pastry and ice cream manufacture (still a significant scheme – disposal measures for butter, butter oil and cream covered a total quantity of 600,000 tonnes of butter equivalents in 2004);
- SMP for use in animal feed;

- skimmed milk for the manufacture of casein/caseinates;
- school milk;
- aid in the form of dairy products for the most deprived people.

Private storage aid

For butter and certain cheeses (mainly Italian cheeses), cheese producers can obtain financial support (aid) for storage costs. Due to seasonal variations in raw milk deliveries the production of some products is high for a short period, which can destabilise markets. This aid stabilises prices by helping producers to take the product temporarily off the market. In the case of butter it also serves as an alternative to intervention.

Milk quotas

The milk quota regime has brought stability to the EU's dairy sector since its introduction in 1984. The CAP reform of 2003 decided that there will be three annual increases of 0.5% of quota volumes for 11 of the EU-15 member states beginning in 2006 (Greece, Ireland and North Ireland, Italy and Spain are the exceptions as they benefit from earlier quota increases). To meet growing demand for dairy products, the European Union decided in March 2008 to increase its milk quotas by 2% beginning in April 2008.

The EU-15 was extended in 2004 with 10 new member states. The national quotas for all EU member states in 2004, including the new countries, are illustrated in Figure 1 (Malta and Cyprus not listed). It is clear that Poland, as new EU-country, belongs to the large dairy countries in Europe, ranking 6th in quota amount of the 25 EU-countries.

In 2007, Romania and Bulgaria entered the EU. The situation in those countries is described later in this paper.

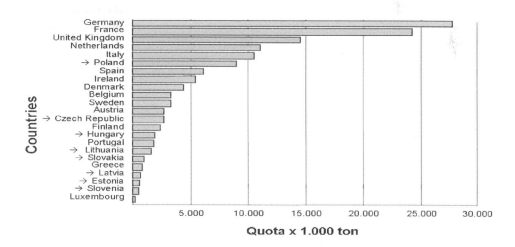

Figure 1. Quota assigned to EU-member states in 2004 (→ indicates the new EU-states; Malta and Cyprus are not listed).

Using trade instruments

Exports

As the EU market price is higher than the world price for dairy products, exports generally take place with the aid of export subsidies. Following the 1994 multilateral trade agreement (known as the Uruguay Round) of the World Trade Organization (WTO), export subsidies have been restricted – both the quantities exported and the amount of total subsidies paid out are strictly limited (Table 1). In practice only cheese exports have reached the quantitative limit each year. Subsidised exports of other dairy products have been well below the volume constraints. The European Commission introduced a tendering system for export refunds on butter, butter oil and SMP in bulk in 2004. This system runs alongside the traditional fixed refund arrangements for all products and has reinforced the more market-oriented approach of the dairy regime post-2003.

Imports

The EU maintains relatively high tariffs on dairy products, in order to sustain the EU market price. There are only minimal imports at full tariff. However, many of the EU's trading partners benefit from special import arrangements – known as Tariff Rate Quotas (TRQs) – whereby imports can come in at lower tariffs. Some of the TRQs are specific to particular exporting countries; others are open to all under the most-favoured nation (MFN) system[1]. TRQs are not always filled (i.e. fully utilised). Those for powders (about 70,000 tonnes) are hardly used; there are TRQs for several different cheese types – amounting to just over 122,000 tones – the average fill rate is 40%; the butter TRQs of approximately 89,000 tonnes are always filled.

Table 1. Subsidised exports – maximum allowable quantities and values for EU-25.

Products	Quantities (tonnes)	Values (€1,000)
Butter/butter oil	399,300	947,800
SMP	272,500	275,800
Cheese	321,300	341,700
Other	958,100	697,700

Making direct payments to farmers

Direct payments or income support 2004-2007 was linked to kg of milk quota. The intention was that the direct income support would be nearly equal to the reduction in intervention prices from 2004 to 2007. Income support was in 2007 decoupled from kg of milk (i.e. from production), the so-called 'decoupling'. Then, the support will be linked to land (ha) or to the farmer. To be eligible for subsidy, the farmer has to comply with a number of conditions, the so-called Cross-Compliance conditions. These conditions refer to all kinds of aspects of sustainable farming or also to a set of 'Good Farming Practices'. Good farming practices that are part of the Cross-Compliance conditions relate to:

---

[1] MFN requires that every time a member state improves the benefits it gives to one trading partner, it must give the same treatment to all other WTO members, so that they remain equal.

- environment;
- product quality and hygiene;
- animal welfare;
- nature management programmes (like preventing erosion and deterioration of biodiversity).

These 'good farming practices' are a result of the various EU-directives, like the Hygiene directive and the Environmental regulations. In some countries, public agencies or dairy companies have incorporated the 'practices' in so-called 'Quality Assurance Schemes'. Farmers are stimulated to participate in such programs.

Characteristics of quota system in relation to structural development

The introduction of a quota system, as happened in 1984 in the 'old' EU-countries and in 2004 in the 'new' EU-countries, is a very complicated process. We have two types of quota: quota for delivering milk to the dairy companies, i.e. to purchasers (Quota A) and quota for direct sales from the farm (Quota D). Introduction of such a system requires institution building, setting up administrative procedures, choices about the system, the choice of priority groups, the handling of the butterfat reference, control aspects, farm management and cost aspects, communication with farmers, and last but not least structural development aspects. In fact the quota system affects the dairy industry and rural development as a whole. This is described for the Eastern European countries by Kuipers *et al.* (2007).

The implementation of the quota system allows different options. The choice of options is very important for the development possibilities of the dairy sector and the individual farm. Especially the way quota transfer is arranged, the built-up of a national reserve and the division of the country in regions or not is essential in this context. The various choices to be made are presented below.

Reference year

The reference year is the year on which the individual quota allocation to farmers is based. For the new member states, this was usually 2002, 2003 or 2004 or a combination of these years. Each member state has been assigned a national quota by the EU (see Figure 1). The national quota is distributed over the national reserve and the individual producers. In some states, the national quota is first divided between regions and/or between the milk purchasers, and in a second step assigned to the individual producers by regional authorities and/or by the purchasers of milk (cooperatives or processing plants). The national quota and individual quota that have been assigned are affected by the development of the cow population and milk volume in the past. The decrease in the number of cattle and corresponding production volume in the Central and Eastern European countries in years 1990-2002 is illustrated in Figure 2.

National reserve

A national reserve is needed to help farmers in specific problem situations and / or to stimulate certain national policies on structural development. Thus the national reserve is meant to:
- give problem farmers a stronger base;
- provide for structural changes.

It is possible to have a more liberal or a more social approach to distributing quota. With a liberal approach the quota transfers are left to the market. With a social approach, the national authorities collect quota at national or regional level and have guidelines for the distribution of this quota. In this case, the quota system is used as a development tool for the dairy sector and for rural development.

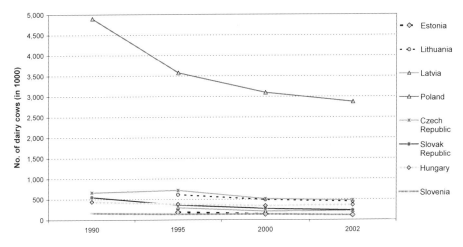

Figure 2. Development of number of cows in new EU-member states.

Possible conditions for providing this additional quota from the national reserve are:
- farmers with development plans;
- young farmers;
- regions with land reform plans.

Quota transfer and structural changes

Different policies which exist concerning quota transfer are as follows:
- Quota Exchange Bureau;
- free transfer of quota;
- transfer of quota linked to land;
- leasing.

The reality is that each country adopts a system that is more or less fitted to the own situation. What can we learn from the systems practised in the Western-European countries? We learn that free quota transfer gives flexibility. Structural developments are not blocked. But when quota prices increase, the cost level of farms also increases. This implies, that the start value of the farm enterprise is higher than without quota. This is not favourable for restructuring and investment in goods other than milk quota.

Also, herd size will affect structural changes. Table 2 shows that herd sizes differ significantly between countries. When herd size is small it may be expected that many small farms will stop milking, i.e. are not allowed to deliver milk to the dairy plants. This is caused by milk quality and economic issues. For the dairy sector, it is an advantage that quota goes free of charge to other farmers. That will keep investment in the sector limited and cost levels competitive. Quota transfer without money transactions are only possible when free transfer is not allowed. In this case all free quota flows to the national reserve. The government Agencies have to distribute this quota among the farmers that remain in milk production. Also, whole farm transfer, like in France, is an option for dairy countries with a small farm structure.

Table 2. Number and size of farms with quota in some countries (2004).

Country	Farms	Cow/farm	Milk prod/cow
Netherlands	30,000	52	7,500
Lithuania	130,000	2.5	5,015
Poland	450,000[1]	4.5	3,840
Hungary	30,000	12	6,317
Estonia	7,120	16	5,119
Czech Republic	3,400	212	5,718
Slovenia	10,900	10.3	4,993

[1] Farms who received quota for delivering to the market; total number of farms was above 700,000.

Regions or not

The philosophy behind the establishment of regions is that quota is forced to stay within each region. It protects certain regions from loosing quota and farms. This is especially the case for the less favoured agricultural regions. The countries that operate regional quota systems are:
- Germany: 21 regions (quota price: 0.20-0.90 €/kg milk)
- France: 3-4 regions
- Poland: 16 provinces

Each region has it's own characteristics. For example, the quota prices of the various regions in Germany in 2004 varied from €0.20-0.90. The lowest prices exist in the former East-Germany regions, and the highest prices are in southern Germany (Bayern). A first indication in 2004 of quota price in Poland was € 0.20 per kg, and in Slovenia € 0.15-0.30 per kg. Also the numbers of quota transfers in the various regions sometimes differ significantly, indicating that more or less restructuring of the dairy sector is going on in different parts of a country. The advantages and disadvantages of instalment of regions are summarised as:
- *Advantage:* protection of certain regions to maintain dairy husbandry in those areas.
- *Disadvantage:* economic developments restricted to the borders of a region.

Present quota situation in new EU countries

The present allocation of milk quota to the EU countries is shown in Figure 1 and the development in the number of dairy farmers with quota is shown in Table 3. In the 'old' EU-countries a reduction in farm numbers of 40-80% has occurred in the period 1995-2007. In the period of 2005-2007 an annual reduction in dairy farm numbers of 3-14% was observed. Especially in Spain, Portugal and Greece the number of farms decreased dramatically. The new countries of Lithuania, Estonia and Poland have also shown a very large decrease in dairy farm numbers in recent years (Table 3).

The various EU countries show a very different structure in size of dairy farms (see Figure 3). The new countries show even greater diversity in this respect. For instance, Slovakia has the largest farm size, while Lithuania has the smallest farms in the EU-25.

The dynamics of dairying in a country is to some degree indicated by the pressure of the farmers on the national quota. Is there a tendency to exceed the individual and national quota or is the national quota not filled up by the farmers? This situation is illustrated respectively for the EU as a whole (Figure 4) and for the individual countries (Figure 5).

As can be seen in Figure 5, there has been a tendency in recent years (2004-2007) that more and more Western European countries did not utilise the national quota. But the situation in the new member States is more complex and difficult to predict. In the first years after entering the EU the

Table 3. Development of number of dairy farmers with a quota in EU (EC, 2007b).

Member state	1995	2005	2007	Change 1995-2007	Annual change 1995-2005	Annual change 2005/2007
Austria	83,793[a]	53,713	47,378	-43.5%	-4.3%	-6.1%
Belgium	24,047	14,533	12,672	-47.3%	-4.9%	-6.6%
Denmark	15,301	6,540	5,354	-65.0%	-8.1%	-9.5%
Germany	230,125	113,020	103,480	-55.0%	-6.9%	-4.3%
Greece	30,316	7,752	6,288	-79.3%	-12.7%	-9.9%
Finland	31,872[a]	17,833	15,213	-52.3%	-5.6%	-7.6%
France	167,593	109,822	100,853	-39.8%	-4.1%	-4.2%
Ireland	48,013	24,194	21,875	-54.4%	-6.6%	-4.9%
Italy	107,011	52,674	46,651	-56.4%	-6.8%	-5.9%
Luxembourg	1,465	991	923	-37.0%	-3.8%	-3.5%
Netherlands	42,249	23,187	21,209	-49.8%	-5.8%	-4.4%
Portugal	73,197	15,804	12,294	-83.2%	-14.2%	-11.8%
Spain	132,352	35,906	28,465	-78.5%	-12.2%	-11.0%
Sweden	17,023[a]	9,449	8,369	-50.8%	-5.7%	-5.9%
UK	41,132	20,629	18,326	-55.4%	-6.7%	-5.7%
Czech Republic		2,991	2,727			-4.5%
Estonia		1,859	1,506			-10.0%
Latvia		25,457	22,141			-6.7%
Lithuania		111,097	82,281			-13.9%
Hungary		6,076	6,175			0.8%
Poland		343,000	276,508			-9.8%
Slovenia		10,578	8,897			-7.9%
Slovakia		814	734			-5.0%

[a] Figures 1996-97.

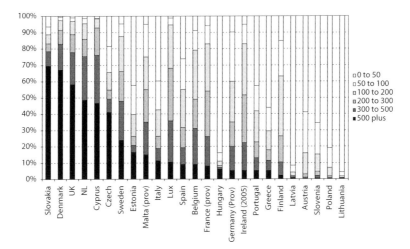

Figure 3. Structure of milk production in EU-25 - distribution of dairy holdings by quota size (tonnes) in year 2006-2007 (EC, 2007b).

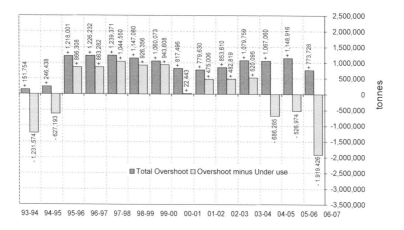

Figure 4. Overshoot and under-use of EU- quota amount in years 1993-1994 till 2006-2007 (EC, 2007b).

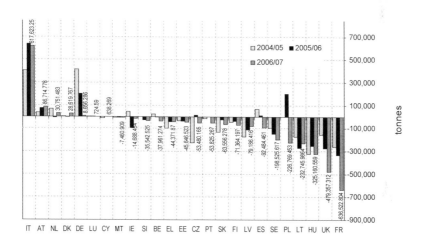

Figure 5. Overshoot or under-use of national quota amounts in EU-25 in years 2004 till 2007 (EC, 2007b).

new member States did not fully utilise their quota. But will this trend continue after adaptation to the new EU-situation? This question will also be tackled in the short descriptions of the dairy situation in the individual countries in the next section.

The dynamics in dairying is also related to the quota prices farmers are prepared to pay. An estimate of the trend in these prices is depicted in Table 4. While in Western European countries, like the Netherlands and Belgium with traditionally high prices, a deceasing trend is visible, in several other countries quota prices are rising. The increase in prices in Denmark and Ireland is particularly noticeable.

When describing the situation in the new countries in the next section, the focus will be on the structural development aspects.

Table 4. Prices paid per kg of quota in different member states in 2007/2008 (Euros).

Member state	2007 market price	Administrative price	Development since last year
The Netherlands	70-80 cents		Decreasing
Belgium (FL/W)		37/25 cents	Decreasing
Finland	6-36 cents	4 cents	Decreasing
Czech Republic	7 cents		Decreasing
Slovenia			Decreasing
Poland	10-20 cents		Decreasing
France		0/15 cents	Stable
Sweden	9 cents		Stable
Austria	50-70 cents		Stable
Germany	23/42 cents		Stable
Spain		27 cents	Stable
Italy	30 cents		Stable
Luxembourg	€1.20		Increasing
United Kingdom	6 cents		Increasing
Denmark	62 cents		Increasing
Ireland	10-28 cents	12 cents	Increasing
Cyprus	€1.33		Increasing
Latvia	43-72 cents		Increasing
Hungary	6 cents		Increasing
Slovakia			No price

Source: Member Sates estimates.

Quota and structural developments in new member states

In this section, we describe the situation in the 10 new EU countries, except Malta and Cyprus. The focus will be on the structural development aspects. This will be done more extensively for five countries, namely Poland, Slovenia, Lithuania, Romania and Bulgaria. Poland is chosen because it is the largest dairy country that entered the EU in 2004. It has very small farms. Slovenia has a mountainous landscape and also relatively small farms. In both countries a form of regionalisation is used in applying the quota system. This was also a reason behind the choice of these two countries. Also, the regionalisation is applied more (Poland) or less (Slovenia) strictly. Lithuania is the country with the smallest average farm size in the EU and has a national operating quota system. Romania and Bulgaria are included, because these countries have recently entered the EU. This way we cover in somewhat more detail a wide spectrum of the existing diversity in Central and Eastern Europe.

Slovenia

General overview of the Slovenian dairy sector

In quota year 2006/2007, the total quota allocated to the Slovenian dairy farming sector was 576,638 tonnes. Out of this, 549,428 tonnes was for milk delivery to dairies (A-quota) and 27,210 was for direct sales (Quota D); see Table 5. Part of the national quota was left as a national reserve for solving problem cases, such as mistakes in allocation or already initiated investment in milk production. During the quota year 2006/2007, a transfer of quota was made: 4,174 tonnes of Quota D was transferred to Quota A, and 126 tonnes from Quota A was moved to Quota D. At the end of quota year 2006/2007, the total quota allocated to Slovenian farmers was 553,476 tonnes for milk delivery to dairies (A quota) and 23,162 tonnes for direct sales (Table 5).

The structure of the dairy sector can be described by the development in number of farms and quota size of farms and the type of farmland (flat, hilly, mountain or less favoured area). The number of

Table 5. Quota for deliveries to dairies and quota for direct sales, 2006/2007 (Agency for Agricultural Market and Rural Development, 2007).

Deliveries (tonnes) – Quota A		Direct sales (tonnes) – Quota D		Total (A+D)
Nat. Ref. Quantity	549,428	Nat. Ref. Quantity	27,210	576,638
Number of producers	8,897	Number of producers	2,320	9,369
Allocated quota	553,476	Allocated quota	23,162	576,638
National Reserve	1,117	National Reserve	250	
Nat. Ref. Quantity on the end of 2006/2007	554,593	Nat. Ref. Quantity on the end of 2006/2007	22,045	576,638

Table 6. Number of dairy farms, dairy cows and quantities of milk sold off farm in Slovenia.

Year	No. of herds	No. of dairy cows	Quantity of milk sold (kg)			No. of dairy cows/farm
			Total	Per cow	Per herd	
1980	55,533	150,694	303,831,000	2,016	5,471	2.7
1985	58,194	175,696	352,454,200	2,120	6,063	2.9
1990	43,656	161,992	359,184,200	2,217	8,228	3.5
1995	30,040	132,532	388,394,400	2,968	12,942	4.4
2000	16,869	117,775	447,831,000	3,758	26,516	6.8
2002	12,589	113,599	473,500,000	4,154	38,577	9.3
2003	11,500	112,484	484,200,000	4,323	42,104	9.7
2004	10,900	112,500	488,683,000	4,344	44,833	10.3
2005	10,578	111,424	506,888,419	4,549	47,919	10.5
2006	9,509	111,000	512,034,328	4,613	53,847	11.7
2007	8,897	106,000	528,426,472	4,985	59,394	11.9

Table 7. Farm structure based on quota for deliveries in the year 2005-2006 (Agency for Agricultural Market and Rural Development, 2006).

Quota size category (tonnes)	Number of farms per quota size category	% of total farms	Quota allocated to farms in this quota size category (tonnes)	% of allocated quota
0-50	6,614	68.77	137,813	26.58
50-100	1,755	18.25	123,699	23.86
100-150	611	6.35	74,049	14.28
150-200	261	2.71	44,946	8.67
200-300	221	2.30	52,501	10.13
300-500	122	1.27	46,069	8.89
500-750	23	0.24	13,531	2.61
>750	10	0.10	25,795	4.98
Total	9,617	100.00	518,404	100.00

Table 8. Farm structure based on deliveries for the year 2007-2008 (Agency for Agricultural Market and Rural Development, 2008).

Quota size category (tonnes)	Number of farms per quota size category	% of total farms	Quota allocated to farms in this quota size category (tonnes)	% of allocated quota
0-50	5,863	64.37	125,302	22.46
50-100	1,786	19.61	126,716	22.71
100-150	699	7.67	84,765	15.19
150-200	308	3.38	52,952	9.49
200-300	259	2.84	62,100	11.13
300-500	146	1.60	55,591	9.96
500-750	35	0.38	20,797	3.73
>750	12	0.13	29,696	5.32
Total	9,108	100.00	557,918	100.00

farms and milk sales is described in Table 6. The quota size of farms in 2005-2006 and 2007-2008 is described in Tables 7 and 8, respectively.

As can be seen, 50% of the national quota in 2005-2006, and 45% of the quota in 2007-2008, belong to farms with less than 100,000 kg. Obviously, there is a small scale dairy structure in Slovenia. In last few years, the percentages of quota in all categories above 100,000 kg are increasing by 0.2-0.5% per year.

Regions

In Slovenia, 12 statistical regions exist. The average size of the farms in the various regions is illustrated in Figure 6. Both deliveries to the purchaser (Quota A) as well as direct sales (Quota D) are presented in Figure 6.

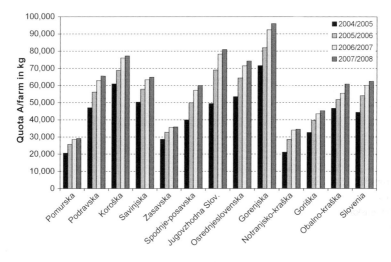

Figure 6. Structure of Slovenian farms regarding the average quota size in 12 regions.

Quota transfer and structural developments

In Slovenia, no official regional quota scheme exists, like in Poland, Germany and France. Nevertheless, the quota transfer rules have been designed in such a way that a structural policy is envisaged. The policy is to protect dairy farming in the hilly and mountain regions.

Within the 12 statistical regions, quota amounts can be transferred from one farm to another without any deduction in quota where quota *is transferred together with land*. The maximum of quota that can be transferred together with land is 15,000 kg of milk/ha. In case that one farmer (farmer A) rents land from another farmer/owner, it is possible to transfer quota from owner of land to farmer A for the duration of the rental period.

Transfer of Quota without Land: in the case of transfer of quota without land, part of the individual quota is allocated to the national reserve, as follows (Republic of Slovenia, 2006):

a. transfer of quota inside statistical region: % to National reserve
- − from farm to farm in the hilly and mountain region   0%
- − from farm to farm inside the regions with limited possibilities for farming (OMD)   5%
- − from farm in lowland to farm with limited possibilities for farming (OMD)   5%
- − from farm to farm in lowland   10%
- − from farm with limited possibilities for farming (OMD) to farm in lowland   10%

b. transfer of quota between statistical regions: % to National reserve
- − from farm to farm in the hilly and mountain regions   0%
- − from farm to farm in regions with limited possibilities for farming (OMD)   25%
- − from farm in lowland to farm with limited possibilities for farming (OMD)   25%
- − from farm to farm in lowland   30%
- − from farm with limited possibilities for farming (OMD) to farm in lowland   30%

When quota is transferred between regions, 25 to 30% of the transferred quota goes to the national reserve, which discourages quota movement from one region to another. However, quota can be transferred without deduction to hilly and mountainous regions.

Did this policy have any effect? In Table 9, we can see the change of quota in the various regions over the last four years. The increase in total quota is due to the enlargement of quota by EU and by the transfers of Quota D (direct sales) to Quota A (deliveries to dairies).

Table 9. Distribution of quota for deliveries to dairies between regions (kg).

Region	Quota A 2004/2005	Quota A 2005/2006	Quota A 2006/2007	Quota A 2007/2008	Index 2006-2007 / 2004-2005
Pomurska	48,778,782	50,242,512	53,520,748	50,554,361	109.7
Podravska	93,160,150	97,418,174	103,412,443	103,018,557	111.0
Koroška	31,075,782	32,663,161	35,110,992	34,983,212	113.0
Savinjska	78,268,847	82,224,073	87,087,225	86,865,253	111.3
Zasavska	3,412,588	3,607,568	3,813,578	3,693,365	111.8
Spodnjeposavska	14,815,907	15,410,827	16,360,068	16,191,470	110.4
Jugovzhodna	50,780,500	52,839,254	56,325,108	55,471,471	110.9
Osrednjeslovenska	79,832,858	82,263,993	86,986,767	85,583,485	109.0
Gorenjska	77,634,231	80,537,654	85,747,185	85,651,799	110.5
Notranjsko-kraška	4,003,636	4,504,829	5,101,046	4,899,338	127.4
Goriška	16,332,622	16,603,997	17,018,971	16,395,299	104.2
Obalno-kraška	687,473	725,155	720,213	670,136	104.8
Total	498,783,376	519,041,197	551,204,344	543,977,746	110.5

The lowest increase in quota is visible in the western part of Slovenia (Goriška and Obalno-krašska regions). These regions belong largely to the less favoured areas. Goriška has also partly very hilly countryside.

In Figures 7 and 8 the overshoot and under-use of quota in these 12 regions is depicted. This indicates that until now most of the regions still have reserve. Four regions (Zasavska, Spodnjeposavska, Gorenjska and Notranjsko-Kraška) have 'no reserve' in milk quota A. The hilly and mountainous areas are partly located in the Gorenjska, Goriška, Koroška and Osrednje-Slovenska regions. Except in the Gorenjska region, these regions still under-use their Quota A.

Each region has its own characteristics. The structural development per region is illustrated in Figure 9. The total amount of milk sold to dairies, the average quota per farm, and the reduction in number of farms during the first 4 quota years, are depicted. A reduction in farm numbers of 12 to 34% has taken place, showing that large differences in regional restructuring exist. The average

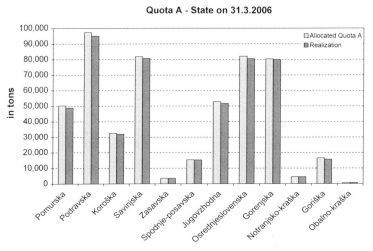

Figure 7. Overshoot and under-use of Quota A in the12 regions in Slovenia.

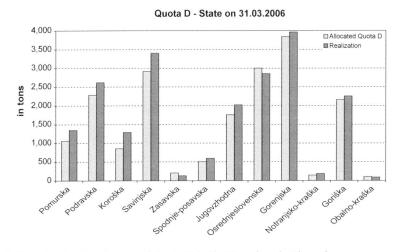

Figure 8. Overshoot and under-use of Quota D in the12 regions in Slovenia.

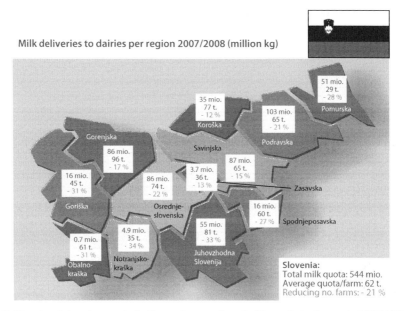

Milk deliveries to dairies per region 2007/2008 (million kg)

Slovenia:
Total milk quota: 544 mio.
Average quota/farm: 62 t.
Reducing no. farms: - 21 %

Figure 9. Structural developments in the various regions in Slovenia in the period 2004/2005 until 2007/2008.

farm size does not seem to be a major factor in this development. Other natural characteristics of the region are apparently more important in affecting the developments.

The market for milk quota has not yet become very active in Slovenia. The quota price in quota year 2006/2007 was between €0.083 and €0.17 per kg milk quota. Such a quota price was influenced by the fact that owners of milk quota received at the end of quota year 2006/2007, the historical rights for their milk premium. This stimulated farmers to buy quota. With regards to the distribution of quota from National Reserve in recent years, in quota year 2007/2008 nearly 'no market' existed for milk quota. For this reason, the price of milk quota for the time being is between €0.00 and €0.05 per kg. However, the price differs between regions. The highest quota price, and also the highest demand for quota, are in the Gorenjska, Savinjska and Osrednje-Slovenska regions. In general we can say that milk quota transfer takes place from the Pomurska, Zasavska, Goriška and Obalno-kraška regions to the Gorenjska, Savinjska and Osrednje-Slovenska regions.

Transfer of quota on a national scale is as follows:

• In quota year 2005/2006 quota was transferred as follows: 8,323,719 kg of Quota A (minus deduction for National Reserve in amount of 598,461 kg of Quota A). It means, that on average 7.2% went to National Reserve. Only 1.7% from total Quota A was transferred from one farm to another farm.

• In quota year 2006/2007 quota was transferred as follows: 5,593,160 kg of Quota A (minus deduction for National Reserve in amount of 376,980 kg of Quota A). It means, that on average 6.7% went to National Reserve. Only 1.0% from total Quota A was transferred from one farm to another farm.

• In quota year 2007/2008 quota was transferred as follows: 3,660,103 kg of Quota A (minus deduction for National Reserve in amount of 252,103 kg of Quota A). It means, that on average 6.9% went to National Reserve. Only 0.7% from total Quota A was transferred from one farm to another farm.

Purchasers of milk

The volume of milk going to the various purchasers has changed dramatically in recent years, especially because of the increasing quantity of milk sold to Italian dairy plants (see Table 10).

Table 10. Sales of milk to Slovenia and Italy plants (litres).

Year	Sale of milk to Slovenian dairy plants	Sale of milk to Italian dairy plants	Fat, %	Protein, %
1980	303,831,000		-	-
1985	352,454,200		-	-
1990	359,184,200		3.74	-
1995	388,394,400		3.93	3.24
2000	447,831,000		4.10	3.36
2002	473,500,000		4.13	3.33
2003	484,200,000		4.14	3.34
2004	486,000,000	2,683,000	4.16	3.36
2005	448,600,000	58,288,419	4.15	3.36
2006	378,129,000	133,905,328	4.09	3.33
2007	338,715,000	189,711,472	4.11	3.34

Poland

General overview of the Polish dairy sector

Poland is a significant dairy producer in Europe with a total milk production of 11.7 million tonnes (Table 11), which places it on the 6th position in Europe. The dairy sector belongs to the most important sectors in the Polish agriculture and food economy. It accounts together with beef production for 26% of Polish agricultural output value. In 2002 there were about 875,000 dairy farms from which about 42% - 376,000 were delivering milk to the processors. In 2007, 657,000 dairy farms were counted of which 247,000 (40%) were delivering milk to the processors. 27,500 farms did have direct sales to the consumer. In April 2007 there were 232 recognised purchasers of milk (J. Falkowski, personal information, 2007).

Change of the economical system and drastic adjustment to the market conditions during the transition period in the 90-ties caused a 43% decline in the dairy herd in period 1990-2005. An increase of the milk efficiency /milk yield per cow (29% during the period 1989-2005) couldn't cover the herd decline what caused an overall cut in milk production of 27% in the same period.

Structure of dairy production in Poland

The dairy sector in Poland characterises a rather high level of diffusion, not comparable with any other 'old' EU Member State. The average statistical dairy herd on a farm in 2005 accounted for 3.93 heads. During the transition process (since 1989), directly connected with radical changes of the economical conditions of production, the average herd size increased only with about 14% (2.9 in 1989 to 3.3 in 2002), while the dairy herd declined with ca 40% (from ca. 5 million in 1989 to

Table 11. Characteristics of milk production in Poland in period 1989-2007 (IERiGŻ, 1990-2007; GUS, 1989-2007).

	1989	1990	1994	1998	2000	2001	2002	2003	2004	2005	2006	2007
Number of dairy cows [1,000 heads]	4,994	4,919	3,863	3,471	3,098	3,005	2,873	2,897	2,796	2,795	2,824	2,787
Index %	100	98.5	77.4	69.5	62.0	60.2	57.5	58.0	56.0	56.0	56.5	55.8
Milk yields [litres/cow/year]	3,260	3,151	3,121	3,491	3,668	3,828	3,902	3,969	4,083	4,200	4,200	4,300
Index %	100	96.7	95.7	107.1	112.5	117.4	119.7	121.7	125.2	128.8	128.8	131.9
Milk production [million litres]	15,926	15,371	11,866	12,178	11,494	11,538	11,527	11,546	11,478	11,600	11,633	11,750
Index %	100	96.5	74.5	76.5	72.2	72.4	72.4	72.5	72.1	72.8	73.0	73.8
Milk deliveries [million litres]	11,385	9,829	6,269	7,070	6,583	7,025	7,219	7,316	7,997	8831	8,419	8,380
Share of deliveries in total milk production %	71.5	63.9	52.8	58.1	57.3	60.9	63.2	63.4	69.7	76.1	72.4	70.9

3 million in 2001) and milk production also with about 30% (from 16 million tons in 1989 to 11.6 million tons in 2005).

During the EU pre-accession period the restructuring of the dairy sector in Poland accelerated. The execution of EU standards (especially sanitary and veterinary norms, and milk quality requirements) as well as the implementation of the CAP instruments (mainly preparations to implement the milk quota system), stimulated producers to start modernising their processes and increasing their scale of production. Investments, financed by farmers' own sources, loans granted by banks and dairy processing enterprises and pre-accession support resulted in an outstanding improvement in milk quality. In the period 1999-2005 the share of extra-class milk (according to the EU standards) in total milk deliveries increased from 35% to 92%. For dairies with an EU certificate this share was even higher, and accounted for 98% of milk deliveries. These strict quality requirements also brought about negative social consequences, however. A lot of mainly small, inefficient producers were not able to adjust, and were forced to either quit milk production or change to semi-subsistence farming. In effect, in 2004 there were 712,000 farms with dairy cows, but only about 48% of them were delivering milk or milk products to the market (Wilkin *et al.*, 2007).

The structure of the dairy sector in Poland is illustrated in Figure 10. In the period 1996-2002 the number of dairy cows kept in the smallest farms (1-9 cows) decreased by 37%, which was mainly due to the fact that 34% of those farms resigned from milk production. Hence, its share in total number of dairy cows diminished from 86% in 1996 to 64% in 2002. At the same time the share of dairy cows kept in middle size farms (10-49) tripled from 8% to 29% (see Figure 10). During 2003-2005, the process of farm concentration further accelerated. The number of dairy farms decreased by 19% whereas dairy cow number shrunk only by 2.7%. In 2005 the share of farms with more then 10 cows in milk sales exceeded 50%, whereas its share in the total dairy cow herd accounted for 40% (Wilkin *et al.*, 2007).

The large changes in milk volume per region depicted in Figure 11 can be explained by the fact that in the period January-July 2004 no quota system yet existed in Poland. So large changes in regional production occurred when the quota system came into sight.

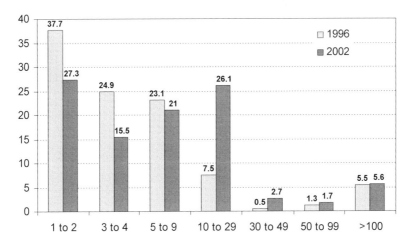

Figure 10. Number of dairy cows and number of dairy farms according to the farming size in 1996 and 2002 (Wilkin *et al.*, 2007).

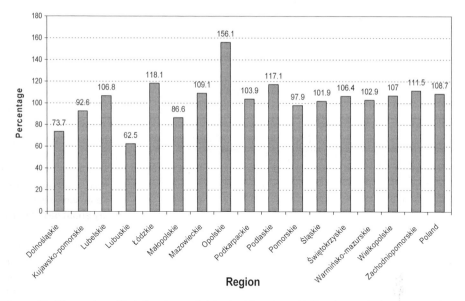

Figure 11. Change in milk deliveries to dairies in 16 regions (delivering period January-July 2005 compared in percentage per region to January-July 2004).

Organisation of the milk quota system and transfer

The model of the quota regime in Poland was based on a regional organisation. The National Authority responsible for implementing and administrating the system is the Agricultural Market Agency (AMA) with support of its 16 regional branches (one in every Province or region). The following ways of transfer are allowed between farms and regions: permanent transfer (to buy/sell/donation), temporary transfer by leasing (in/out) and conversion of the wholesale/direct sale quota (permanently or temporary).

- *Permanent transfer regarding the regional organisation of the system.* Permanent transfer of quota is allowed only between the producers who have their farm in the same region of RB AMA (however, in 2007 is decided that the regional limits of quota transfer will be abolished as of beginning of the 2009/2010 milk quota year).
- *Temporary contracts – leasing of quota.* The leasing contract is valid until the end of the current milk quota year. Leasing transfers are allowed only within the wholesale producers with farms located in the same region of RB AMA.

In the final pre-accession negotiations, Poland received 8.96 million tonnes of milk quota from which 95.4% as a wholesale quota (limit for deliveries to processing). In addition, it was assigned 0.426 million tonnes for the restructuring reserve to be used as of 2005/2006. The first quota allocation has been made at the beginning of 2004, directly to producers with respect to their deliveries during the reference year (1 April 2002 – 31 March 2003). According to data of the Agricultural Market Agency, during the first allocation ca. 355 thousand producers received the wholesale limit and 78 thousand producers the direct sales quota. Wholesale quota assigned in 2004/2005 was ca. 1.2 million tonnes (14%) higher than deliveries to processing in 2003, hence there was some space for development in milk deliveries. Nevertheless, due to a dynamic restructuring of milk production during the first milk quota year, this gap was almost fully supplied by the producers.

Deliveries to processing during the first milk quota year (2004/2005) were ca. 2% lower than wholesale quota assigned. At the same time direct deliveries to consumers were 33.4% lower than direct sales quota, hence total national quota was not binding the production. Moreover, according to accession treaty, the first year of the milk quota system in Poland (2004/2005) was not charged by the super-levy payment. However, due to a very dynamic development of marketed milk production in Poland, deliveries to processing during the first milk quota year (2004/2005) increased by 13.5%, (Figure 11) accompanied at the same time by a reduction of direct deliveries to consumers and self supplies on the farm. As a result of this restructuring, the total milk production in Poland has not changed and the commercial production ratio increased. Development in marketed milk production has been followed by a farm concentration process. During the first milk quota year number of registered wholesale producers decreased (Figure 12) by 12.6% and direct sale farms by 40%. Consequently, the average wholesale quota assigned to the farms increased by ca. 30% to the level of 27 thousand kg.

In the second milk quota year 2005/2006 milk deliveries increased by a further 5%, what in consequence contributed to the national quota overrun by 1.79% (ca. 287.6 thousand tonnes)[2]. The Polish surplus was third largest in the EU after Italy and Germany. None of the 'new member' Countries, except Czech Republic, has exceeded the national quota. During the first two years of the milk quota system functioning in Poland (April 2004 - April 2006), the number of dairy producers delivering milk to processing decreased by 21.4% (Figure 11). At the same time, milk deliveries raised and hence the average milk production per commercial farm has grown by 47%.

Although above changes affected the whole country, significant differences between regions with respect to rate and scope of restructuring have been observed. There are 5 regions from the 16 administrative regions in Poland with a large and increasing production, accounting in 2006 for 67% of milk delivered to dairies. These are: Mazowieckie, Podlaskie, Wielkopolskie, Warmińsko-Mazurskie and Łódzkie. It might be observed that milk production moves towards the northern part of Poland, where very good natural conditions, a long tradition of milk economy and favourable agrarian structure have facilitated its development. These regions characterise also a high number of large herds (over 10 cows), where lives ca 50-60% of the total cow population.

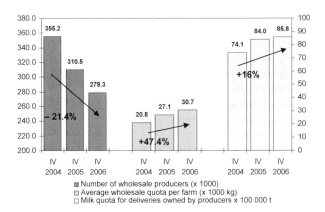

Figure 12. Characteristics of milk wholesale deliveries at the beginning of the 2004/2005, 2005/2006, 2006/2007 milk quota years (Own calculations basing on data provided by Agricultural Market Agency).

---

[2] Wholesale quota was oversupplied by 3.32%, but direct sale quota was undersupplied by 39.38%, hence in effect total excess reached 1.79% (data provided by Agricultural Market Agency, August 2006).

The milk supply limit in the third quota year (2006/2007) was enlarged by the restructuring reserve, what increased initially the assigned limit by 5%. After this change the entire Polish milk quota amounted 9,379 mld kg. As it was expected the national milk quota in 2006/2007 and 2007/2008 were not overrun.

Influence of the quota transfer system on structural development

Since the milk quota system is binding production to the area, because transfer of the milk quota is not permitted between the regions, milk production restructuring has been disturbed. The main problems concerning milk quotas are two: first, the plafond which is limiting production and was oversupplied in 2005/2006 (so producers were punished by the levy payment), and secondly, the quota trade restrictions, which result in rather high quota prices and inhibit restructuring of milk production. The point is that quota is allowed to be traded only between the farmers having their holding in the same administrative region. That rule is extremely unfavourable for restructuring of dairy production. In the regions where milk production developed very fast, like in Podlaskie, Wielkopolskie and Mazowieckie, a very high demand for production quotas is observed. This demand can not be met by existing supply, therefore quota prices in these regions are a few times higher than in other regions. Producers who want to develop their production have to bear a very high costs of investments or pay the super-levy for overrunning the quota. These transfer procedures create also a huge barrier for processing companies, which have to search for the raw milk outside the region, increasing transaction costs. This rule inhibits transition of quota from the less favourable milk production regions (mainly south of Poland) to those with good conditions for milk production. However, this unfavourable rule will be abandoned at the beginning of the 2009 milk quota year. Despite of all constrains concerning milk quota, the milk production sector in Poland has been developing in recent years in a fast speed: rigorous structural developments took place within the various regions. These changes are depicted in Figure 13. The average farm quota size per region

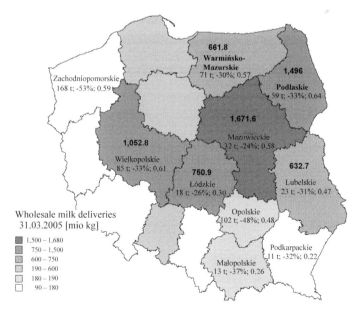

Figure 13. Structural developments in the various regions in Poland in period 2004/2005 till 2007/2008. The numbers concern: average quota amount/farm in tonnes (2007/2008); % reduction in no. of farms in 4 year period; price of quota/kg in Zlote (2008).

varied from 11 tonnes (in Podkarpackie region) to 168 tonnes (in Zachodniopomorskie region). The reduction in number of farms with quota in last 4 quota years fluctuates from 24% (in Mazowieckie region) to 53% (in Zachodniopomorskie region). Also the quota price is listed in Figure 13: it varies from 0,06 euro per kg milk in Podkaropackie region till 0,17 euro per kg in the Podlaski region.

There appeared to be some relationships between the parameters measured in the various regions, which are used to illustrate structural developments. Especially the average farm size of a region, expressed in average quota amount per farm per region, is a strong indicator. For instance, the correlation between the average quota farm size and the reduction in number of farms in a certain region in last 4 years appeared to be r=0.70, indicating that a larger farm size goes together with a stronger restructuring process. A larger average farm quota size in a certain region is also positively correlated (r=0.60) with a tendency to exceed the individual quota in that region more often. Apparently, larger farms in Poland seem to have a greater pressure to produce. A larger average farm quota size also relates to a higher quota price in that region (r=0.51). In other words, larger farms have more pressure to produce and too enlarge the farm and, therefore, are prepared to pay on average more per kg quota bought.

Even though the quota was enlarged by restructuring reserve and enlarged by a further 2% in the 2008 milk quota year, it will continue to hamper farm development in forthcoming years. Together with growing milk yields, the dairy cow number will have to be gradually reduced to keep up the production within the restricted level. Growing internal demand with slightly reduced export, while production is limited, may cause import growth At the same time the farm concentration process will be necessary to compete on the EU market and to deal with the growing production costs. On the other hand, quota trade will be freed from the region limits in 2009, which will create more incentives for regional concentration of milk production and will enhance the process of milk production restructuring further. It can be expected that the process of policy-driven restructuring will continue in the near future, since new regulations require additional investments from either the processing or the milk producing sector. Policy driven restructuring will be also caused by reform of the EU dairy market organisation resulting from both: future debate on the Common Agricultural Policy as well as WTO negotiations.

Romania

As a candidate EU member state, Romania obtained a milk quota allocation from the European Commission by the end of 2004 and is fixed at 3,245,000 tonnes of which:
• 1,093,000 tonnes for deliveries to processors;
• 1,964,000 tonnes for direct sales;
• 188,000 tonnes of national reserve.
The reference year for the fat production is 2004. The reference period for volume is 1 April 2005 - 31 March 2006.

Evaluation on request for milk quota system by 1 April 2007

By April 1 2007, a milk quota system was in place and since then allocations have been made to those who applied for, and are eligible to, obtain milk quota. However, due to many circumstances, allocation of quota is still under discussion and will be subject to changes. On the basis of the applications made, the following conclusions can be drawn:
• Delivery quota
   − national allocation 1,093,000 tonnes;
   − of which in reserve 21,860 tonnes;
   − available for allocation 1,071,140 tonnes;

- request for allocation 1,022,607 tonnes;
- 2% is made available 1,002,155 tonnes;
- 68,985 tonnes is reserved for farms with extensions of the farm (e.g. land) and new investments in the farm (e.g. housing).

250,493 holdings (farms) applied for quota for direct delivery to milk processing facilities. These requests can be divided in the following quantities as shown in Table 12. In terms of farm size, the request for deliveries to processors are shown in Table 13.

- Direct sales quota
  - national allocation 1,964,000 tonnes;
  - of which reserve 39,280 tonnes;
  - available for allocation 1,924,720 tonnes;
  - request for allocation 2,186,431 tonnes.

622,504 holdings (farms/individuals) applied for quota for direct sales. These requests can be divided in the following quantities (Table 14). Only four farms have a quota > 1 million kg.

- Allocation of quota for extensions and new holdings
  - available for extensions and new holdings 68,985 tonnes;
  - number of applications 1,233;
  - quantity of milk requested 181,588 tonnes;

Table 12. Volume requests by farms for direct delivery to milk processing facilities.

Volume (kg)	%
<5,000	40.4
5,000-10,000	16.4
10,001-50,000	17.8
50,001-100,000	19.9
>100,000	5.8

Table 13. Farms size requesting direct delivery to milk processing facilities.

Number of cows per applicant	%
<2	43.6
3 to 5	18.0
6 to 10	7.9
11 to 50	16.4
51 to 100	3.8
>101	10.3

Table 14. Volume requests by farms for direct sales.

Volume (kg)	%
<5,000	66.7
5,000-10,000	14.9
10,001-50,000	13.0
50,001–1,000,000	5.1
>1,000,000	0.4

- estimation of the allocation 38%.
- Quota transfer and regions
  In Romania, 8 regions are established, each with a separate quota. Only transfers within regions are allowed. Within a region, free transfer of quota exists. However, 20% of each transfer is taken away to go to the national reserve. Therefore, regionalisation and a considerable redistribution of quota are envisaged in Romania

Bulgaria

The dairy sector reform turned out to be a much more serious challenge for the local industry and policy makers than initially thought (USDA, 2007; A. Kuipers, personal communication). This report states that 'the reform had lots of economic, social and political implications. Politicians had to take difficult decisions, not always based on the best market approach. Lack of trust between various private and public players further complicate reform implementation'.

Dairy sector development

Restructuring of the dairy sector has continued in 2006 and 2007 with further commercialisation, consolidation and enlargement of bigger farms, and decline in the number and role of smaller family-type farms. In 2006/2007, there has been a significant growth in the number of farms with 10-20 cows (32%) and farms with 20-30 cows (15%). Similarly, the number of farms with more than 100 dairy cows rose 18% although they still account for only 7% of all cows. Farms with only one cow (backyard rural semi-subsistence farms) were 10% less in 2006 than in 2005, and accounted for 25% of Bulgaria's dairy cows. The number of small farms raising up to 9 dairy cows has fallen by 8% (see Table 15).

According to the National Dairy Board (NDB) data, as of March 2007, Bulgaria had 96,595 registered dairy farms, which are grouped in three categories depending on their compliance with the EU hygiene and milk quality standards:
- First category (fully corresponding to EU standards) – 1,125 farms.
- Second category (meeting EU equipment and hygiene standards but not fully meeting milk quality standards) – 1,238 farms.
- Third category (not meeting EU standards) – 94,232 farms.

Out of the total registered farms 90,415 are 'small' with up to 10 cows, 5,749 are 'average' size with 10-50 cows, and 431 are 'large' farms (>50 cows). According to another industry source (Association of Milk Processors (AMP)) 78% of farmers have 1-9 cows. The milk is collected at 4,200 milk collection stations.

Table 15. Structure of the dairy farm sector in Bulgaria in 2006 (MinAg, 2007).

Dairy cows per a farm	Farms		Dairy cows	
	Number	Change 2005/2006	Number	Change 2005/2006
1-2	113,328	-9.9%	140,500	-9.6%
3-9	21,470	0.5%	92,400	2.6%
10-19	3,552	31.8%	44,700	28.1%
20 and more	1,578	2.5%	72,500	7.9%
Total	139,928	-7.5%	350,100	0.7%

Milk supply

The total amount of milk produced in 2006 was 1,515 million tonnes of which 1,298 million tonnes (86%) cow milk, 107,000 million tonnes (7%) sheep milk and 102,000 million tonnes (6,7%) goat milk. The 2006 milk supply was slightly (0.5%) higher than in 2005, with a stable growth in buffalo and sheep milk, a decline in goat milk and a slight increase in cow milk. The average milk yield in 2006 was 3,600 litres per cow, 1,428 litres per buffalo; 87.3 litres per ewe and 221 litres per goat (Table 16). The production of processed dairy products was 12.6% less than in 2005. About 50% of all fresh milk produced on farms was delivered to dairies.

Introduction and distribution of milk quotas in May-July 2007 caused lots of economic and political turmoil. The dairy industry was split over the issue. The two major industry organisations, the NDB and AMP, expressed very different positions. These differences were conceptual, sometimes politicised, or related to certain economic interests. The NDB has 8 regional dairy boards. It is supportive of faster market reforms and concentration despite the negative social/political effects. The AMP has 120 members. It is more concerned about the social and political effects in rural areas and the goal is to continue production of smaller dairy farms as a backbone of the dairy industry.

The national reference quantity for milk set for Bulgaria in 2007/2008 is 979,000 tonnes, of which 722,000 tonnes are for deliveries and 257,000 tonnes are for direct sales, much less than the traditional production of 1.2-1.3 million tonnes. A 'reserve' quota of 39,180 tonnes may be added in 2009 (counting current on-farm consumption). The reference average fat content is 3.91%.

As of June/July 2007, a total of 96,572 farmers had individual dairy quotas. The average dairy farm has a quota of 7.0 tonnes for deliveries, and 3.0 tonnes for direct sales. According to the NDB, 5,100 farms have more than 10 cows (5.5% of all dairy farms) and can produce under quota 220,000 tonnes of milk. The number of large farms (more than 50 cows) is 446 and their milk quota is in total 189,000 tonnes. These figures are disputable, according to processors, who claim that only 35% of the total milk quota is produced by the large farms, while the remaining 65% is collected from small family type farms.

Milk deliveries quota

Table 16. Production of milk at farms by type of dairy livestock for 2002-2006 in Bulgaria in 2006 (tonnes) (MinAg, 2007).

Year	Cow	Buffalo	Sheep	Goat	Total
2002	1,305,912	4,410	93,479	104,820	1,508,621
2003	1,308,525	5,276	88,679	101,530	1,504,010
2004	1,344,750	6,229	117,682	129,381	1,598,042
2005	1,286,909	6,989	105,057	109,114	1,508,069
2006	1,298,709	7,132	107,535	102,297	1,515,673
2006 in 1000 liters	1,260,883	6,891	104,201	99,414	1,471,389
Change 2005 vs. 2006	0.9%	2.0%	2.4%	-6.2%	0.5%

As the demand for milk deliveries exceeded the quota by more than 200,000 million tonnes (NDB data) the NDB had to reduce the quantity of requested milk to the size of the quota. This reduction was applied mainly to farms that produce sub-standard quality milk (the third category, see above), based on methodology approved by the Ministry of Agriculture and Forestry in Decree #51.

A special reserve within the milk deliveries of 92,463 tonnes was set based on MinAg information about investment projects. This decision was approved by the Agricultural Minister in Ordinance 09-231/ April 13, 2007.

Thus, the quantity of milk deliveries that remained for actual distribution was 630,000 tonnes. It was distributed as follows: 192,400 tonnes to all farms in the first category (1,125) meeting 100% of their applications, 94,600 tonnes to all farms in the second category (1,238) also meeting 100% of their applications, and 342,500 tonnes to farms in the third category (94,209), which is 47% less than their applications. By regions, the regional dairy boards in Plovdiv and Rousse received the highest shares, 22% and 17%, respectively.

Reduction in quotas for the third category farms, the most numerous group, caused protests. Although there is clear legislation (Decree #51) about quota distribution, the lack of a public register of farms and their individual quotas resulted in speculations about NDB justification for its decision. On some of the reduced quota farms, the quota is lower than the average milk yield. For example, in the Yambol area, there are 3,606 dairy farms in the third category compared to 48 in the first, and 123 farms in the second, categories. Most of these farms will not be able to go to the second category at the end of 2007, which means that in 2008 many will be shut down and dairy cattle will be either slaughtered or sold. In region of Dobrich, farmers started to sell cows at a low price of €250 or slaughter them due to reduced quotas (140,000 tonnes requested and 90,000 tonnes approved).

Milk direct sales quota

Milk direct sales quotas did not attract many farm applications – all requests were fully met since farms applied for only 77,600 tonnes. Out of that, 10,213 tonnes were distributed to the first category farms; 6,176 tonnes to the second category, and 61,202 tonnes to the third category. By regions, the highest share of quotas was received by Blagoevgrad (30%) and Sliven, (26%). Thus, 186,000 tonnes remained unused.

Dairy manufacturers response

Currently, dairy manufacturers evaluate their fixed expenses for registration, reporting, monitoring and traceability for milk from smaller farms as much higher than their profit from milk processing. Many will try to optimise milk purchases by switching to a smaller number of larger suppliers, and/ or imports of powder milk/whey as a substitute for fresh milk. The industry estimates that Bulgarian milk production will be efficient only when most farms produce 6-7,000 liters average milk yield with a potential for 10,000 liters (currently, it is about 3,600 liters) which means not more than 170,000-200,000 cows to meet the current production ceiling of 979,000 tonnes.

Dairy products market

The issue of milk quality is tightly related to the dynamism of the dairy products market as follows:
- Commercial supply of major dairy products (liquid milk, yogurt and cheese) has slowly increased since 2003 at the expense of home production although it still remains high.
- Over the past 5 years, distribution and sales of dairy products have steadily moved from traditional smaller retail outlets to supermarkets and hypermarkets where quality and hygiene requirements are more stringent. For example, in 2006/2007, 48% to 89% of various types of liquid milk were sold in super/hypermarkets.
- Retail quality and safety requirements for dairy products become increasingly stringent. In early July, major retailers announced that they would require a new certification IFS (International Food

Standard, introduced by German and French retailers, recently accepted by Italy) standards from local suppliers. Since only a few have such a standard in place, the pressure for more investment is likely to increase, which in turn will affect the purchases of raw milk. If Bulgarian companies cannot respond fast enough to retailers' demand, imports from other EU countries can quickly replace local supply.

The dairy processing sector continues to attract foreign investors: two of the largest dairy companies were purchased by foreign companies. As an example, the dairy company Fama is the market leader in North-East Bulgaria and it is located close to the Romanian market. It has a well developed milk purchasing and collection system, well known brands, experience in working under a private label and has good distribution networks (capacity is 180 tonnes milk for daily processing). The ice cream market also attracted several major investors over the past year. Currently, the market size is estimated at €40 million or 11,000 m t with the prospect to grow to 15,000 m t or 20% in the next 1-2 years. Top companies on this market are Nestle (via Delta) 35% market share, and Darko with 22% market share. The remaining 45% is split among several local companies with about 10% share each (Karil, Izida, Deni). In 2006, the companies registered a growth of 5-20% compared to 2005, with the trend continuing in 2007.

Quota transfer and regions

Bulgaria is divided in 8 quota regions to start with. So regionalisation is envisaged. The rules concerning quota transfers are still in development.

Slovakia

In Slovakia in 2004, 85% of the national herd of 211,000 cows were managed on 820 farms. Indeed, Slovakia has large farms. The remainder of cows are on very small farms, mainly for subsistence farming. This kind of farming exists in the Northern Slovakian regions.

Slovakia does not have a free market for milk quota. Quota is fixed to the farm, i.e. to the cows. When a farmer sells cows, his quota is decreased; when he purchases cows, his quota is increased without paying for the quota. A whole farm with quota can be transferred to another farmer. This is a similar system as in France.

There is not a regional quota system. The filling of the national quota in Slovakia from year 2004-2007 was between 93% and 95%.

Hungary

Hungary is a special country in respect to the quota system. It was the only country joining the EU in 2004 that did have a national quota system in place before entering. The quota system was established in 1996. The situation is also unique, because Hungary has a mixture of very large and small farms. In 2004, about 1000 large farms had on average about 600 cows. In addition, more than 20,000 small farms had about 7 cows on average.

The total quantity of national quota is 1,947,280 tonnes, of which 1,782,650 tonnes are for deliveries and 164,630 tonnes is for direct sales.

Transfer of quota was initially restricted. Hungary is considered as one region.

Czech Republic

The number of dairy cattle in Czech Republic decreased by 37% from 1989-2003 (1.248 million to 0.46 million). In 2004, 43% of the farms had between 100 and 300 cows, and 38% of farmers had more than 300 cows. The Czech Republic, Slovakia and Hungary (partly) have the largest farms in Europe.

There is no regional quota system and transfer of quota is free.

Baltic countries

In the Baltic countries, no quota regions are established, thus free transfer of quota between farms exists. Also leasing of quota is allowed. The administration of quota and the application of the quota rules are very liberal in these countries.

The restructuring is very strong, especially in Estonia. In the Baltic countries, non-use of agricultural land has become normal indicating drastic changes in agricultural structures.

In Estonia, 2,428 farmers were assigned a quota in 2004. However, 4,356 farmers did not apply for quota. These smaller farms obviously will leave the sector in years to come. About 57% of the quota holders have 1-10 cows, 29% have 10-50 cows, 4% have 50-100 cows, 9% have 100-500 cows, and 1% have above 500 cows. This is a remarkable distribution with a large spread from small to large. Nearly 10% of dairy farms have from 100-500 cows. The structure of farming in Estonia differs very much from those in Lithuania and Latvia.

Lithuania had 448,000 dairy cows in 2004. These cows are managed on 270.635 farms, which indicate an average of about 2 cows per farm.

In Latvia, there were 4,800 cows kept in 2004 giving the average herd size of nearly 6.

Case study of extra attention to the structural developments in Lithuania: auction of quota

Lithuania is the only country in the EU with an auction system of transfer of quota. Sellers and buyers meet each other via internet at an auction. This system is somewhat comparable to the Quota Exchange Bureaus in Denmark (centralised Bureau) and Germany (at regional level). Up until the present, 3 auctions have been held. The amount of quota exchanged and prices paid are listed in Table 17.

Structural developments

The average quota of the farms in the various regions in Lithuania varies from 7.8 tonnes (Svenčionys region) to 30.6 tonnes (Siauliai region). This illustrates the very small dairy farm size in Lithuania. A rigorous restructuring of the dairy sector is taking place (see Figure 14). In the last 4 quota years, there was a reduction in the number of farms with quota of 31% (Tauragé region) to 46% (Rokiškis region). This restructuring looks somewhat similar to the situation in Poland (see Figure 13), while the changes in Slovenia show a much more modest trend (see Figure 9).

It would be very interesting to study the factors behind the large differences in restructuring between the countries but also between the various regions in the different countries.

Table 17. Quota exchange in Lithuania through an auction system (1 euro = 3.413 litters).

	1st auction (November 2007)	2nd auction (March 2008)	3rd auction (November 2008)
No. of potential sellers	2,119	2,489	933
Total amount for sale in 1000 tonnes	19.7	23.5	11.5
No. of potential buyers	328	380	134
Total amount requested in 1000 tonnes	11.7	10.7	3.7
No. of buyers	244	342	127
Total amount purchased in 1000 tonnes	4.8	9.6	3.7
No. of sellers	576	842	251
Price sellers demanded			
Average in litters/kg	0.46	0.22	0.11
Lowest versus highest	0.05-2.50	0.05-1.50	0.01-2.0
Price buyers paid			
Average in litters/kg	0.28	0.20	0.06
Lowest versus highest	0.23-0.60	0.10-0.50	0.03-0.20

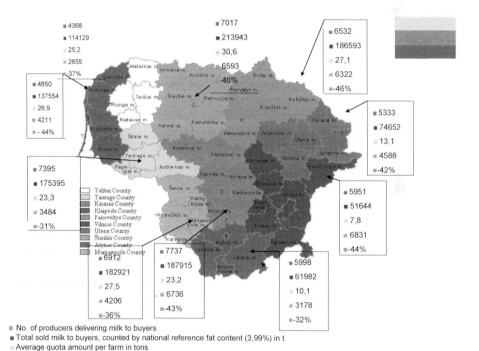

No. of producers delivering milk to buyers
Total sold milk to buyers, counted by national reference fat content (3,99%) in t
Average quota amount per farm in tons
Reduction in no of farms with quota from 2004/2005 to 2007/2008
Percentage reduction in no. of farms from 2004 to 2008

Figure 14. Structural developments in the various regions in Lithuania in the period from 2004/2005 to 2007/2008.

References

Agency for Agricultural Market and Rural Development, 2008. Sodobno Kmetijstvo, Yearbook 41, March 2008.

Agency for Agricultural Market and Rural Development, 2007. Sodobno Kmetijstvo, Yearbook 40, No. 1, March 2007.

Agency for Agricultural Market and Rural Development, 2006. Special Supplement 'Sodobno Kmetijstvo', Yearbook 39, No. 2, May 2006.

European Commission (EC), 2007a. Milk and milk products in the European Union. Available at: http://ec.europa.eu/agriculture/publi/fact/milk/2007_en.pdf.

European Commission (EC), 2007b. Report of Commission to Council, Brussels 12-12-2007 (COM-2007-800final).

GUS, various years. Statistical yearbooks (volumes from 1989-2007). GUS (Main Statistical Office), Warszawa.

IERiGZ, various years. Rynek mleka stan i perspektywy. Market Analyses Series (volumes from 1990-2007), Warszawa ISSN 1231-2673.

Kuipers, A., Klopcic, M. and Svitojus, A. (eds.), 2006. Farm management and extension needs in Central and Eastern European Countries under the EU milk quota. EAAP Technical Series, No.8, Wageningen Academic Publishers, the Netherlands.

MinAg, 2007. Ministry of Agriculture statistical bulletin #105. Ministry of Agriculture, Bulgaria.

Republic of Slovenia, 2006. UREDBA. o spremembah in dopolnitvah Uredbe o uvedbi dajatve za mleko in mlečne proizvode. Uradni list RS, št. 132/2006 z dne 15.12.2006, str. 14588 (Official gazette of the Republic of Slovenia, No. 132/06, published 16.12.2006, p. 14588).

USDA, 2007. Bulgaria dairy and products - a cup of hot milk for Bulgarian dairy reform 2007. Gain report number BU7023. Date 7/23/2007. United States Department of Agriculture, Foreign Agricultural Service. Available at: www.fas.usda.gov/gainfiles/200707/146291846.doc

Wilkin, J., Milczarek, D., Malak-Rawlikowska, A. and Fałkowski, J., 2007. The dairy sector in Poland. Regoverning markets agrifood sector study, IIED, London.

The well-being of Heifer International animals

T.S. Wollen and D.P. Bhandari

Heifer International, 1 World Avenue, Little Rock, Arkansas 72202, USA; terry.wollen@heifer.org

Abstract

Heifer International is a non-governmental organisation (NGO) providing livestock and training in countries of Central and Eastern Europe with offices in 10 of those countries. Heifer projects are formed around strong local community groups and provide quality livestock, training and related support to men, women and youth in order to assist with food security and to improve livelihoods. Training and extension services are provided to assist families to improve animal housing, management, breeding, nutrition and veterinary health. Animals receive humane handling and protection and in turn provide food, other by-products and work to the family. Training in agro-ecology integrates livestock production into sustainable farming systems, which protect and enhance the local environment. Groups are taught how animals can become a vital part of family activities without causing a burden on farm resources and are integrated into the community. Animal well-being recommendations are provided according to the topical areas of animal nutrition, animal reproduction, animal health, animal husbandry, care for the environmental and humane slaughter. Each section lists critical elements to achieve in order to enhance the health and production of the animals. Even though relatively simple, the community group can use these recommendations for three important tasks. They form an outline of items to include in the preparation of animal health-related proposals to funders. They also form the training program outline for livestock participants. Finally, they serve as a tool for monitoring and evaluating the success of projects related to livestock. The overall objective of these recommendations is to select and rear animals that improve family livelihoods and that are suitable to available resources. As Heifer works with trained animal husbandry staff and health care specialists in all locations, these recommendations can be adapted by local livestock project management as needed to local conditions and objectives. This is the case for many resource-deficient areas where livestock keepers lack extensive resources for livestock care and often manage animal husbandry at basic levels. Further on the spectrum of livestock farms are larger scale operations, for which product quality, quantity and uniformity are important for the external market and require a higher level of management. Regardless of the scale, local customs should always be taken into consideration including traditional practices, local leadership and other experiences.

Keywords: animal well-being, animal nutrition, animal reproduction, animal health, animal husbandry, care for the environment, humane slaughter

Introduction

Heifer International is a non-governmental organisation (NGO) providing livestock and training in 18 countries of Central and Eastern Europe with offices in 10 of those countries: Albania, Armenia, Bulgaria, Kosovo, Lithuania, Poland, Romania, Russia, Slovakia and Ukraine.

To ensure the appropriate care of Heifer project animals, all recipient families and communities participate in extensive training prior to receiving their 'living loans'. Through training and extension services, Heifer enables families to improve animal housing, management, breeding, nutrition and veterinary health (Aaker, 2007: 124). All animals receive humane handling and protection while generating food and income for families engaged in the programs. Training in agroecology allows recipients to integrate livestock production into sustainable farming systems, which protect and enhance the local environment. In this manner, animals receive adequate care, improve farm conditions and become a vital part of a family livelihood.

Training in these practices is provided by Heifer technical staff or project partners, who themselves have been trained in Heifer's model of participatory methods of values-based community development (Aaker, 2007: 46). This model is based on lessons from the field, including an understanding of the local situation and practices, envisioning opportunities and obstacles to success, designing action plans and follow-up monitoring and evaluation. Local animal health and husbandry specialists are enlisted to provide specific lessons on veterinary and agro-ecological subjects.

Animal well-being recommendations

Heifer International has developed *animal well-being recommendations* to serve as guidelines for livestock recipients and community-based animal health workers (Heifer International, 2008). Heifer's recommendations encompass more than just animal health, encouraging holistic animal management. This holistic approach includes all facets of animal husbandry, nutrition, reproduction, humane slaughter and environmental impact – reinforced by continuous learning and monitoring of success.

The overall objective of the following *recommendations* is to select and raise animals that improve family livelihoods and that are suitable to the resources available in the respective community. As Heifer works with trained animal husbandry staff and health care specialists in all locations, these standard *recommendations* can be modified by local livestock project management as needed to adapt to local conditions and objectives. This is the case for many resource-deficient areas where livestock keepers lack extensive external resources for livestock care and can thus manage animal health and husbandry at basic levels. On the opposite end of this spectrum are for-profit livestock operations - for which product quantity, quality and uniformity are important - requiring a higher level of health and husbandry management. Regardless of the scale, local customs should also be taken into consideration for community level management, including the use of indigenous practices, traditional leadership and other experiences.

Selection of appropriate animal management for Heifer International livestock will be made from the six major areas of consideration. Each contain numerous recommendations for improved husbandry and indicators that assist the livestock manager in monitoring his / her management progress.

Animal nutrition

Animal nutrition is one of the major concerns for Heifer in field project activities. Project beneficiaries are encouraged to provide nutrition that is appropriate to age, gestation stage, and production and growth requirements. There should be a continuous supply fresh water to all animals.

Feed must be stored to retain quality and to safeguard it from contamination and rodents. Individual feed ingredients can be combined with local forages to meet livestock, poultry and fish nutrition requirements. Fodder, energy and or protein reserves should be stored for times when fresh feeds are not available. Pasture grasses can be organised and warehoused for nearly year-round feeding, depending on the climate. Intensively managed rotational grazing gives farmers more benefits.

For producers following National Organic Standards, certified organic feed substances must be stored according to regulations and pastures maintained according to the rules with proper records maintained.

Animal reproduction

Animal reproduction management includes appropriate selection procedures of females and males of each animal species, well-designed breeding programs, recorded gestation calendars and birthing care of animals that Heifer provides to community members.

Animals are selected for genetic improvement based on recipient community's economic conditions and ability to feed and provide care. Pregnancy of female livestock is certified prior to purchase. The Heifer community group is involved in the selection process to guarantee that project participants have husbandry skills, available resources and a market for the offspring and other products. Local animal types and breeds that are suitable for the local environment are the best choice for reproduction, when available and in good quality.

A breeding plan is prepared with the appropriate male breeding stock for natural breeding or artificial insemination. Breeding plans are organised for all reproductively active animals. Appropriate equipment and a trained technician are used when employing artificial insemination. Appropriate care for male animals is provided during non-breeding season.

During gestation, feed and care for pregnant animals is provided in a proper manner: pregnant females are not overused for draft-power and adequate nutrition levels are maintained according to stage of gestation.

At the time of birth, a dry, heated (when necessary) area is provided that is protected from severe weather as much as possible and appropriate for the respective livestock species. Afterward, appropriate nutrition through lactation is provided to insure timely rebreeding. Birthing cycles are tracked, recorded and monitored in order to meet species capability under the local conditions.

Animal health

Heifer International supports basic preventive health care programs, treatment of disease conditions and surveillance for emerging local, catastrophic and zoonotic diseases.

The use of ethno veterinary products, local healers and traditional remedies when appropriate and available are promoted, especially in very rural and remote areas. Commercial medications are purchased through reliable sources to avoid counterfeits. Records of product names, manufacturer, supplier, lot number and expiration date are kept. A cold chain is provided when required for medicine and vaccines. Medicines will be stored in clean, safe, temperature-controlled environment, when required. Expired products are destroyed immediately and hazardous products are stored according to label recommendations.

All medicines, insecticides, pesticides and other products are used strictly according to the label unless under supervision of veterinary advice. Alternative therapies and management practices will be considered for deworming and use of antibiotics in order to reduce development of resistance. Appropriate delivery devices and protective clothing are utilised when needed. Milk withholding times and slaughter withdrawal times are observed.

Sick and recovering animals are isolated from healthy stock when possible. Mortality will be investigated and diagnosed in order to avoid recurring illness. Keeping a record of all medical and alternative treatments and mortality is good for building future health programs.

Disease prevention is normally achieved by vaccination schedules and strategic deworming programs. All preventive treatments are recorded. Internal and external parasites are monitored, managed and treated. Health checks for mineral sufficiency for livestock on rotational grazing pastures should be established.

Disease surveillance can be an asset to local government livestock departments and diagnostic laboratories. Pre-purchase health checks, blood tests (especially those required by the government) and fecal counts, are good precautions when purchasing new animals for the community. New herd animals are put in quarantine, usually for 30 days. Diagnostic procedures for herd illness are established and diseased or dead animals are disposed of completely. Local authorities will be involved to monitor zoonotic diseases.

Animal husbandry

Appropriate animal management, facilities, hygiene and record keeping are discussed here under these recommendations at a minimum basic level for new livestock recipients.

Animal management: Animals should be identified in a humane and sanitary manner by ear tag, tattoo, brand, ear-notch or other means. Records are kept on each animal from birth or purchase until the animal leaves the farm. Functional and humane restraint equipment (e.g. nose rings) are installed. Animals are kept safe from injury by training staff in improved animal management. Tail docking, teeth clipping, castration, dehorning, and other equipment must be maintain in good working. Tack is kept clean and oiled. The same is true for shearing equipment. Technicians are trained for each respective health and management procedure. Wool and other by-products adequately stored for top market prices and local use.

Animal facilities: Zero grazing units should be built correctly, stocked properly, cleaned regularly and the manure collected for use in farming systems. Animals are given daily exercise and sunlight when kept in confinement. Appropriate fencing is utilised for animal species; kept tight and in good repair. Appropriate gates are utilised and kept them in good working order. Dry shelter and housing is provided with adequate lighting and airflow. Ample lots, pastures and woodlands that are hazard-free and not over-used will be used.

Milk hygiene: Milk facilities and equipment are kept clean and meet local milk ordinance standards for production type. Milkers and milk handlers must practice good personal hygiene. The appropriate milking time for each animal is established and milk handling and cooling is conducted in a timely fashion. Milk product quality protected and product safety promoted through sound milk handling and preservation methods for raw and processed products.

Animal records: Records for pure-bred, indigenous, rare and heritage breed animals are maintained up-to-date. Breeding records are made available such as: breeding dates, sire, birth date, growth, weaning, and body condition scoring data. Health records are maintained including incidents, treatment, vaccination and final disposition. Causes of mortality and disposition of the carcass are documented.

Care for the environment

Heifer International encourages diversification of livestock and crop species to avoid harsh impact on the environment. In addition, manure and pasture management are key factors taken into consideration.

Manure management: Manure is removed regularly and composted in an appropriate location. Manure is utilised for crop fertiliser, bio-gas and other applications. Manure is stored adequately to avoid pests

Pasture management: An adequate pasture / livestock density is ensured to allow for proper animal nutrition with minimal supplements. The land and pastures are managed to control soil erosion. Overconsumption of grasses, shrubs and trees for fodder is avoided to ensure best regeneration.

Animal slaughter

From the animal well-being perspective, Heifer International has developed basic recommendations for humane slaughter of animals for consumption and for sale.

Humane slaughter includes the following practices: (1) slaughter of any animal is only conducted by properly trained individuals; (2) slaughter equipment is kept in sharp / good working order; (3) slaughter is completed as quickly as possible to minimise animal anxiety; (4) only healthy animals are slaughtered for food; and (5) any animal that has died from unknown causes will not be used for human consumption.

Training

Training is another major piece of overall activity for Heifer International at all levels. Each country program should have adequately trained staff that is connected to outside technical expertise. Training will be provided to all project recipients in all aspects of animal well-being. Heifer staff and /or project partners should be equipped with sufficient skills and resources for training and monitoring. Recipients must have sufficient resources and other abilities to utilise the training that is provided. Best practices are documented and shared throughout country programs and beyond. Local culture and traditions are included in all training.

Monitoring

Monitoring is important to guarantee the successful application of each of the aforementioned recommendations. It is therefore essential that the program calendar for trained staff include monitoring time, which should be completed on schedule (Aaker, 2007: 105-116).

Each Heifer International country office should also create an *animal well-being committee* with the task of reviewing and monitoring project conditions on a periodic basis. The committee would serve as the forum of discussion when issues of well-being are presented, hear questions of abuse brought before the group for evaluation and make recommendations to project managers, program partners, and the country director. The committee should include a veterinarian (from the local country Heifer staff, if available), a program staff person and a program partner / lay-person from outside Heifer. The committee should meet regularly and plan a monitoring visit to each project site at least once a year. At times, the committee may deem it necessary to remove an animal from a project family due to unhealthy or abusive conditions. Project contracts should, therefore, contain provisions stipulating what constitutes unhealthy or abusive conditions and the actions that will be taken. Project animals may be subjected to a variety of undesirable conditions that pose no life risk, but are inhumane. A few examples are listed below:

- serious medical problems without adequate veterinary care;
- lack of adequate food or water;
- exposure to extreme temperatures;
- inadequate shelter or bedding;
- housing in unsanitary enclosures.

References

Aaker, J., 2007. The Heifer model, cornerstones; values-based development. A training manual. Heifer International, Little Rock, USA.

Heifer International, Animal Well-Being, 2008. A training document. Heifer International, Little Rock, USA, 8 pp.

Part 2 Country reports

The dairy cattle sector in the Russian Federation shows a tendency to improvement

S. Kharitonov[1], I. Yanchukov[2], A. Ermilov[2], Y. Grigoriev[3] and O. Osadchaya[3]

[1]Russian State Agrarian University – MATA, Timiryazevskaya str. 48, 127550 Moscow, Russia; haritonov@timacad.ru; [2]Moscow Regional Breeding Organization, Vinogradnaya str. 9-Б, Dolgoprudniy, 141706 Moscow region, Russia; [3]All Russian State Research Institute on Animal Husbandry, Dubrovitsy, Podolskiy District, 142132 Moscow Region, Russia

Abstract

The dairy cattle sector is one of the most important branches of animal husbandry in the Russian Federation. In the period from 1992 to 2000, production of milk and milk products decreased significantly. The main reason for this was the unstable situation in the country which resulted in reduced animal numbers and productivity. This negative trend in milk production was halted in 2005 when the National Priority-Project on Development of the Agrarian sector was established and started to be implemented by the Government. This paper outlines the main characteristics of milk production and milk consumption, structure of milk production by federal regions and by various types of farms, breed composition, the organisational scheme of breeding in the dairy cattle population, and perspectives on the development of the dairy sector in the Russian Federation.

Keywords: agricultural enterprises, breed, breeding management, dairy cattle, households, milk production, milk consumption

Introduction

During the 1990s, Russia experienced some dramatic changes in its social-economic structure. The break-up of the Soviet Union and transition of the national economic system from centralised state control to being market oriented had a great impact on the existing equilibrium for different branches of the national economy, their structure, and the influence of the state sector on the general economic situation in the country.

A breach in the price parity for agricultural produce on one hand, and energy sources, farm machinery and equipment, fertilisers and mixed feeds on the other, resulted in a sharp decline in agricultural production and output by agricultural enterprises, which in the main were subject to nationalisation (privatisation and restructuring to joint stock companies). As a result, the gross output of agricultural products decreased considerably. In animal husbandry (i.e. dairy cattle sector), it resulted in a decline in livestock numbers for agricultural enterprises, lower productivity of farm animals, and structural reorganisation of the agrarian sector. If in the early nineties, agricultural enterprises were 100% represented by collective farms and state farms. Already by the mid nineties, the major large and medium size livestock farms were transformed to joint stock companies, agricultural cooperatives and other non-state amalgamations. At the same time, some private farms were developing and they started to produce agricultural goods, not only for their own consumption (as it happened in most cases in individual households of the rural population), but for sale.

Production and consumption of milk products in the Russian Federation

In the livestock industry of the Russian Federation, one of the main sectors that plays a leading role in the production of animal origin products is dairy cattle husbandry.

For the reasons that have been mentioned above, production of milk in all categories of dairy farms in Russia declined very significantly during the period from 1990 to 2005. While in 1990 the production of milk was 55.7 mt, the amount of milk produced in 2005 was only 31.1 mt, a reduction of 44.1% (Table 1). Afterwards, this tendency stopped, and since 2005, the production of milk has slowly increased. According to official information of the Central Statistical Bureau of the Russian Federation, in 2007, total production of milk in all categories of farms was 31.4 m t (Dzaparidze and Khrestin, 2008).

According to official data for the period studied (2005-2007), world milk production increased by 5.0%. The main increase in milk production was observed in the Asian region (11.2%) and the countries of the European Union (5.0%). In the Russian Federation, the tendency to increase milk production was not so strong (about 1.0%), but a stabilisation of milk production in the country is obvious. Data showing the ratio between export and import of milk products in different countries are presented in Table 2.

It is evident that the imports of milk products into the Russian Federation is much greater than the exports. During the period 2005-2007, the country imported 5.1%-5.9% of annually world traded milk products. At the same time, export of milk products from Russia was only 0.2-0.3 mt. Thus, the ratio between import and export of milk products ranged from 1:7-1:13.5. In dynamic aspects from 2005 to 2007, import of milk products decreased by 33%, while export became less by 33%. The same situation is observed in countries of the Asian region where imports of dairy products was 4.8-5.4 times greater than exports.

In contrast to Russia and Asian countries, the states of the European Union and USA exported more milk products than they imported. For instance, export of milk in European countries was 6-7 times

Table 1. Production of milk (Dzaparidze and Khrestin, 2008).

Country/region	Milk produced (mt)		
	2005	2006	2007
World total	642.3	656.8	674.6
European Union	146.9	145.5	154.2
USA	80.3	82.5	83.5
Asia region	216.2	227.8	240.3
Russian Federation	31.1	31.2	31.4

Table 2. Import and export of milk products (Dzaparidze and Khrestin, 2008).

Country/region	Import (mt)			Export (mt)		
	2005	2006	2007	2005	2006	2007
World total	43.6	45.2	45.5	45.4	47.5	48.0
European Union	2.0	2.0	2.0	13.7	12.8	12.7
USA	2.3	2.0	2.0	4.6	4.7	4.8
Asia region	21.8	23.6	23.7	4.5	4.4	4.5
Russian Federation	2.2	2.6	2.7	0.3	0.2	0.2

more than imported. In addition, it needs to be noted that imported dairy products by the Russian Federation comprised 8-9% of total value of milk produced in the country.

Russia takes the 5th place in the world on consumption of milk, ranked behind the countries of the European Union, United States, India and China. The dynamic trend in milk consumption for the last 5 years is presented in Figure 1.

Figure 1 shows that the consumption of milk in the European Union is tending to decline slightly. The level of consumption of dairy products in the EU countries in 2007 comprised 95.5% of the 2003 values. In the same period, the situation with milk products consumed by the population of the US was more stable: 27.2 mt of dairy products was consumed in 2003 compared with 27.4 mt in 2007. The dynamic trend in dairy products consumption in the Russian Federation looks similar to the changes in the European countries, but on another level. While the Russian population consumed 13.4 mt of milk products in 2003, the value of milk products consumption had declined by 10% in 2007. In 2007, the use of milk products in Russia accounted for 43.8% and 35.3%, respectively of the same indices in the United States and European Union.

As for the main categories of milk products, it can be concluded (according official sources) that in the Russian Federation the consumption of cheese has increased from 498,000 t in 2003 to 660,000 t in 2007 (32.5%). Consumption of milk butter in the same period was stable at about 440,000 t per year with a small annual variation (about 5-7%).

According to the norms established by the Russian Academy of Medical Science (RAMS), the provision of milk and milk products (expressed as liquid milk) should be 392 kg per capita. In Table 3, the consumption of milk and main milk products in the Russian Federation and in developed countries is shown. It is obvious that the level of consumption of the main milk products is much less than required to meet the norms of RAMS, and compared to corresponding indices in developed countries.

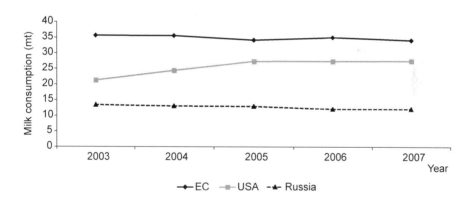

Figure 1. Dynamic trend in milk consumption in selected countries.

Table 3. Provision of milk and milk products (kg per capita) (Dzaparidze and Khrestin, 2008).

Product	Amount required to the norms of RAMS	Consumption	
		Russia	Developed countries
Milk	114	64	112-182
Sour milk drinks	32	14.4	29-46
Cheese	21	6.2	18-29
Milk Butter	6.1	4.2	2.1-7.7

Russia is divided into 7 federal territorial administrative regions, e.g. Central, North-Western, Southern, Privolzhskiy, Ural, Siberian and Far Eastern, which in turn are subdivided into republics, krais and oblasts. The structure of milk production by federal regions is presented in Table 4. The figures in Table 4 show that 86% of total milk was produced in 4 regions: Privolzhskiy (33%), Central (20%), Siberian (17%) and Southern (16%). The rest of the country (3 regions: North-Western, Ural and Far Eastern) together produced just 4.5 mt of milk accounting for 14% of total milk production in Russia.

One of the main reasons that milk production decreased in the Russian Federation was a diminution in dairy cattle numbers. While in 1990 the number of cattle was 57,043,000 head, at the end of 2007 it was only 23,310,000 on all categories of farms. The dynamic changes in the numbers of dairy cows in different types of farms is presented in Table 5.

It is evident that in the period examined the maximum curtailment in number of dairy cows occurred in agricultural enterprises: the number of cows on these farms declined by 39% and amounted 3.97 m in 2007. The same tendency was also observed in the private sector, although it was rather small compared to the agricultural enterprises. As a result, the total population of cows decreased by 26.5% in 8 years. It is remarkably to note that livestock numbers declined in all regions and all oblasts of the Russian Federation.

Table 6 provides information on the productivity of dairy cattle in agricultural enterprises, as well as farmers' households and private auxiliary households.

In the early years of the 20th century, the situation with milk production in the Russian Federation was more or less stable. In spite of a significant reduction in the number of dairy cattle in agricultural enterprises, the total milk produced in this category of farms essentially did not decrease. A stable situation with milk production was observed in the private sector and an absolutely positive tendency has been shown in farmers' households. In contrast with the nineties of last century, when a reduction of dairy cattle population was accompanied by a similar decrease in milk production, the total volume of milk produced in years 2000-2007 in different types of dairy farms has remained the same: 32.3 mt in 2000 and 32.2 mt in 2007.

Table 4. Structure of milk produced by federal regions in 2007 (Dzaparidze and Khrestin, 2008).

Federal region	Milk produced	
	Absolute value (mt)	%
Central	6.44	20
North-Western	1.93	6
Southern	5.15	16
Privolzhskiy	10.63	33
Ural	1.93	6
Siberian	5.47	17
Far Eastern	0.64	2

Table 5. Numbers of dairy cows in various categories of farms (m) (Dzaparidze and Khrestin, 2008).

Category of farm	Year							
	2000	2001	2002	2003	2004	2005	2006	2007
Agricultural enterprises	6.49	6.09	5.65	5.13	4.67	4.29	4.07	3.97
Farmers' households	0.25	0.28	0.31	0.35	0.36	0.41	0.47	0.49
Private sector	5.98	5.85	5.79	5.61	5.22	4.85	4.86	4.85
Total	12.66	12.22	11.75	11.09	10.25	9.55	9.41	9.30

Table 6. Milk output from different types of farms (mt) (Dzaparidze and Khrestin, 2008).

Category of farm	Year							
	2000	2001	2002	2003	2004	2005	2006	2007
Agricultural enterprises	15.3	15.5	16.0	15.4	14.4	14.0	14.1	14.2
Peasants' households	0.6	0.6	0.7	0.8	0.9	1.0	1.1	1.3
Households of the population	16.4	16.8	16.8	17.2	16.9	16.2	16.2	16.7
Total	32.3	32.9	33.5	33.4	32.2	31.2	31.4	32.2

In this context, it needs to be kept in mind that the greatest role in milk production is played by agricultural enterprises and by households of the population, which produced respectively 44.0% and 52.0% of milk in 2007. The proportion of total milk produced in farmers' households was not so substantial and amounted to only 3.9% of total, although it increased by more than two fold (1.8% in 2000 vs. 3.9% in 2007).

The stable situation with production volume in the country has been achieved in spite of a reduction in the dairy livestock population due to rising milk productivity of cows. The data on dynamic trends in average milk yield per cow in agricultural enterprises of various regions is shown in Table 7.

According to Table 7, average milk yield per cow during the period under review has increased from 2,341 kg in 2000 to 3,798 kg in 2007. The increase in average milk yield for this period was 62.2% for the whole of Russia. The highest average productivity per cow occurred in the North-Eastern region (4,753 kg) where milk yield increased by 59.9% from the base of the year 2000. In other federal regions, alterations in average milk yield for the period under review was from 47.9% in the Far-Eastern region to 67.8% in the Southern region.

Table 7. Average milk yield per cow/year in agricultural enterprises (Dzaparidze and Khrestin, 2008).

Region	Average milk yield in year (kg)							
	2000	2001	2002	2003	2004	2005	2006	2007
Central	2,358	2,609	2,798	2,981	3,130	3,319	3,613	3,790
North-Western	2973	3,350	3,637	3,878	4,059	4,295	4,629	4,753
Southern	2,467	2,728	2,968	3,088	3,238	3,669	3,973	4,140
Privolzhskiy	2,277	2,460	2,718	2,924	2,988	3,155	3,463	3,476
Ural	2,359	2,352	2,688	3,029	3,137	3,360	3,793	3,971
Siberian	2,157	2,369	2,690	2,721	2,684	2,983	3,106	3,302
Far Eastern	1,725	1,691	1,884	1,952	1,999	2,047	2,265	2,550
Total	2,341	2,551	2,878	2,979	3,068	3,292	3,623	3,798

Breed composition of dairy cattle of Russia

The population of dairy cattle in the Russian Federation is represented by 24 breeds which can be structurally classified as follows:
- Black and white breeds which include: Black and White, Istobenskaya, Tagilskaya, and Holstein of Black and White color.
- Simmental and related breeds: Simmental, Sychevskaya, Red and White, and Holstein of Red and White color.

- Brown breeds: Brown Swiss and Kostromskaya.
- Red breeds: Red steppe, Suksunskaya, Bestuzhevskaya, Red Gorbatovskaya, and Red Tambovskaya.
- Original breeds: Ayrshire, Kholmogorskaya, Yaroslavskaya, Jersey, Dagestan mountain cattle, Zebu cattle, and Yakutskaya.

It is allowed to use any interclass crosses and offspring shall be considered purebreds of the dam breed. The percentage distribution of different breeds of dairy cattle in the Russian Federation is shown in Table 8.

The data in Table 9 summarise the productivity of the pedigree dairy cattle population. The group of black and white breeds is the most widely spread and it is found in all regions of the Russian Federation. It is one of the few breeds that increased in number of animals from year to year. The common feature of the breeds of this group is that the animals originated from Dutch and North German cattle, and this served as a basis to combine them in one common group. During the previous 40-50 years, animals of the black and white Holstein breed (sires in majority) have been widely used, and are being used now, in this group to increase milk yield and shape, as well as conformation and adaptability of the udder to machine milking. The reproduction of Holstein genetic material goes in two directions: due to the imports of worldwide breeding stock (from USA, Canada, recently from Germany, Netherlands, Denmark, Hungary) and due to the introduction of locally bred sires and their intensive use for breeding purposes. In recent years, embryo transplantation is being widely used, mainly for the production of sires. The black and white, as rule, has the highest potential for milk yield with moderate butter fat and protein contents.

At present, Russia is using a breed-regional system of dairy cattle selection. Of course it doesn't mean that any regional population is presented as an isolated group but exchange of genetic material

Table 8. Percentage distribution of livestock breeds of dairy cattle (Shapochkin, 2007).

Breed (group of breeds)	Relative number of animals, %			Relative number of cows, %		
	2005	2006	difference	2005	2006	difference
Black and White	55.2	56.5	+1.3	54.9	56.4	+1.5
Simmental	13.4	12.6	-0.8	13.2	12.2	-1.0
Kholmogorskaya	9.2	9.6	+0.4	9.5	9.8	+0.3
Red steppe	5.7	5.0	-0.7	5.7	5.2	-0.5
Red and White	3.7	3.7	0	3.5	3.6	+0.1
Others (19)	12.8	12.6	-0.2	13.2	12.8	-0.4
	100%	100%		100%	100%	

Table 9. Average productive traits of cows in the main breeds of dairy cattle (breeding part of population) (Shapochkin, 2007).

Breed	Milk yield (kg)			Fat content (%)		
	2005	2006	difference	2005	2006	difference
Black and White	4,209	4,483	+274	3.75	3.75	0
Simmental	3,138	3,307	+169	3.73	3.76	+0.03
Kholmogorskaya	3,791	3,972	+181	3.68	3.71	+0.03
Red steppe	3,592	3,819	+227	3.82	3.83	+0.01
Red and White	4,004	4,222	+218	3.80	3.83	+0.03
All breeds	3,937	4,190	+253	3.76	3.77	+0.01

is restricted by local interests. The regional breeding systems are based on the oblast programs of selection targeted to the general improvement of productivity and breeding performance of animals. The Leningrad and Moscow regions as well as the Sverdlovsk, Perm and Omsk regions take the leading positions in raising black and white dairy cattle.

Until the 1960s, Simmental and related breeds dominated in Russia. In these years, the population of Simmental cattle in the Black soil zone reached 95% of the total number of dairy cattle. At that time the largest herd size (over 800,000 head) was in the Voronezh region.

By the nineties, the Simmental population in many traditional rearing zones had decreased to less than half (Samara, Kaluga, Tula oblasts) and had nearly disappeared in Kemerovo, Amur oblasts and in Khabarovsk and Primorskiy krais.

In the 1990s, the Red and White breed of dairy cattle was approved in Russia. This breed was selected by using the Red and White dairy cattle in the population of purebred Simmental cattle which was represented as a typical dual-purpose breed. In the group of Simmental related breeds, the Red and White breed is distinguished as significantly specialised dairy cattle with high milk yields and improved udder quality.

The majority of modern cattle breeds classified as Brown Swiss breed, were selected in Switzerland and adjacent mountain regions of Germany, Austria and Italy. Since 1995, the Brown cattle in Russia are subdivided into two breeds: Brown Swiss (which includes a small number of cattle of the Dagestan Brown Caucasian breed as a type because it had no sires at AI stations) and the Kostromskaya breed. The first group of Brown Swiss animals was imported to Russia from Switzerland in 1861 to a farm of the Moscow Agricultural Academy. This breed soon became very popular in the Moscow, Smolensk, Tula and other regions of Russia. Now the Brown breeds have lost their once pre-eminent position in the total dairy cattle breeding population in Russia. Actually, Brown breeds of cattle have almost disappeared in the Moscow region but are still preserved in some farms of Vladimir, Nizhniy Novgorod, Bryansk, Tula oblasts as well in Tatarstan. Now the population of Brown breed cows is about 6.5% of the total number of cattle in Russia and is a tending to further reduce in number of animals.

Red cattle have a long history in Russia, since early 18th century, when the first immigrants from Germany and Holland came to settle in Russia. They brought with them cattle of different Red and Black and white breeds and their predecessors can still be found in Germany, Switzerland, France and Holland. At the end of 19th century, Angler and Dutch breeds were used to genetically improve these cattle. It is worth mentioning that attempts to cross Red Steppe with imported Angler and Red Dutch cattle in the seventies and nineties failed to substantially increase milk yields and now some farms cross red cows with Red and White Holstein sires.

The last group of cattle breeds has a specific significance since each breed is independent and as rule is not used for crossing with other breeds except with the Kholmogorskaya and Yaroslavskaya breeds where introduction of Black and White Holstein genes is allowed. In the sixties and seventies of 20th century, the total number of Kholmogorskaya breed cows exceeded 1.1 million in 35 oblasts of Russia. The number of oblasts has increased due to the regions and republics that have a typical cold climate (Murmansk, Kamchatka, Magadan, Tjumen, Yakutia). In 1993, the Pecherskiy type of breed was confirmed, which is well adapted to the severe climatic conditions of the Northern part of the Komi republic.

The Yaroslavskaya breed was selected in the 20th century in the Central European part of Russia by using local Northern forest cattle. The Kholmogorskaya breed did have a significant influence, on the formation of the existing population of the Yarislavskaya breed, while to a lesser extent, an impact of Black and White Dutch cattle still exists. At present, the Yaroslavskaya breed of dairy cattle is common in Yaroslavl, Ivanovo, Vologda and Tver oblasts of the Russian Federation.

The remaining breeds of dairy cattle, except the Airshire breed, belong to local, practically isolated, breeds raised under certain environmental conditions and well adapted to them. The average productive traits of cows in the main populations of dairy cattle are shown in Table 9. It is clear that the average milk yield as well as fat content have increased significantly in all the main breeds of Russian dairy cattle. It has resulted in an increase in milk yield of 6.4%, and in fat content of 0.002% in the whole breeding population of dairy cattle in Russia.

System of breeding work in dairy cattle of Russia

The management of genetic resources in the Russian Federation is based on the Federal law 'On Pedigree Animal Husbandry'. This legislative document was introduced in 1995. It defines the basic provisions for activities (conditions, requirements, obligations, rights) of breeding animal owners (organisations, enterprises, joint stock companies, farmers' households, private entrepreneurs) irrespective of the type of ownership (Russian Federation, 1995). The farm animal population is divided into three types of pedigree organisations:
• breeding plants which are represented by the best breeding farms;
• breeding reproducers which answer to breeding requirements but have worse results of breeding activities than breeding plants have;
• organisations on artificial insemination of farm animals (AI stations).
The latest official results of the breeding activities for the first two types of pedigree organisations are presented in Table 10.
Data in Table 10 show that compared with 2008, in 2006, the number of breeding farms was considerably lower: for breeding plants by 16 units and for breeding reproducers by 79 units. This has led to a decrease in the number of pedigree animals. In 2006, the proportion of recorded animals was little more than 5% of the total dairy livestock population. On a positive note, the average milk yield per cow in the breeding part of the population significantly exceeded the value for in the whole population - in breeding plants by 75%, and in breeding reproducers by 35%. It was the fundamental reason why economic indices increased in breeding farms. The profitability rate reached 17-19%.
Beside breeding farms and AI stations, the organisational scheme of management of dairy cattle breeding resources in Russia includes two upper levels: federal and regional ones. On the federal level, the Ministry of Agriculture is responsible for elaboration of federal programs aimed at improving the socio-economic situation in the whole rural sector. In animal husbandry, the Federal Ministry works out general concepts, specific standards and principles to support programs implementation. On regional levels, there are state regional administrations which are responsible for implementation of programs in regions (republics, krais, oblasts). They are working in close collaboration with the Federal Ministry, but they are free to introduce their own programs in the framework of their authorities and resources.

Table 10. Main results of breeding work in pedigree enterprises in the dairy cattle sector of Russia (Shapochkin, 2007).

Traits	Breeding plants		Breeding reproducers	
	2006	± to 2005	2006	± to 2005
Number of farms (units)	300	-16	680	-79
Number of animals (thousand heads)	456.1	-10.4	671.3	-39.1
Number of cows (thousand heads)	255.0	-3.0	370.9	-19.3
Average milk yield (kg)	6267	+267	4838	+242
Average fat content (%)	3.88	+0.01	3.81	0
Profitability rate (%)	19.3	+1.0	17.1	+0.7

In conclusion, the present situation of dairy cattle in the Russian Federation has become more stable and predictable than it was in the 1990s, but remains rather complicated. The official document 'Concept-Forecast for Russia's Animal Husbandry Development up to Year 2010', which has been approved by the scientific session of the Russian Academy of Agricultural Sciences and Governing Board of the Russian Ministry of Agriculture, envisages the following actions (Ministry of Agriculture of the Russian Federation, 2001):

- restoration and development of the population and structure as well as preservation of the unique gene stock of breeding animals;
- creation of favorable conditions for investment policy in this sector;
- raising the economic efficiency of activities pursued by breeding organisations and enterprises.

To resolve these issues, it has been decided first of all to improve the normative – methodological as well as economical and material foundation of cattle breeding, which is aimed at:

- increasing the number of breeding herds and animals under registration (identification, maintaining the data base to be used as a basis for official herd books of pedigree animals);
- increasing the pace of genetic progress for breeding animal populations according to the selected characteristics due to the introduction and optimisation of breeding programs with the populations of farm animals;
- optimisation of the breeding organisations infrastructure (associations on breeds, system on farm animals artificial insemination, independent laboratories to register phenotypical characteristics and appraisal of the animals genetic value);
- increasing the effectiveness of distribution of the best genetic resources, its rational use and realisation of the potential under the real conditions of agricultural production;
- conducting the objective monitoring of the breeding livestock sector, projection of its development and optimisation of breeding programs;
- Russia's accession to international organisations dealing with pedigree animal husbandry.

Implementation of these steps into the practice of dairy cattle husbandry in the Russian Federation offers the opportunity to look ahead with definite optimism.

References

Dzaparidze, T. and I. Khrestin (eds.), 2008. Inquiry book of dairy market in Russia. Moscow, Russian Federation, 335 pp.

Shapochkin, V. (ed), 2007. Year-book on results of breeding work in dairy cattle of Russian Federation. Lesnye Polyany, Moscow region, Russian Federation, 287 pp.

Ministry of Agriculture of the Russian Federation, 2001. Concept-forecast for Russia's animal husbandry development up to year 2010. Published as special issue.

Russian Federation, 1995. Federal Law 'On animal breeding', legislative number 123. Approved by State Duma on 12.07.1995 and signed by President of the Russian Federation on 03.08. 1995.

Cattle sector and dairy chain developments in Baltic countries

E. Gedgaudas

Lithuanian Cattle Breeders Association, Kalvarijos g. 128, Kaunas 46403, Lithuania;
edvardasg@litgenas.lt

Abstract

Dairy farming historically and traditionally is still one of the most important agricultural sectors in all three Baltic countries. In the beginning of the 20th century, animal recording was started and breeding associations were established. They were working very actively until collectivisation. When Baltic countries became independent from the Soviet Union, the breeding system had to be re-established in all three countries. Dairy cooperative farms, agricultural companies and family farms started their activities, but the main part was small keepers with 2-3 cows. Animal identification, milk recording, data processing, milk test and breeding processes were developed. Before becoming EU members, Baltic countries received very big support for agriculture from EU funds. It helped to increase farm size and to modernise cattle farms and offered possibilities to improve the quality of herd management. It also created the possibility for increased production, allowed a reduction in the price of milk, facilitated improved animal welfare and achieved EU standards for milk. The Estonian Holstein is the dominant breed in Estonia; the Latvian Brown is the dominant breed in Latvia, and the Lithuanian Black and White is the dominant breed in Lithuania. Cows of different breeds are being milk recorded and average milk yield, protein and fat contents differ between breeds. In all Baltic countries, farmers use more and more Holstein breed semen for insemination of their cows. In the future, Holstein will be the main dairy breed. Total number of cattle in Estonia is 242,000, including 112,700 cows, in Latvia total number of cattle is 389,700, including 195,600 cows, in Lithuania the total cattle number is 787,900, including 414,800 cows. Milk recording is best developed in Estonia, where, in 2007, 90.9% cows were in milk recorded. In Latvia it was 70%, and in Lithuania it was only 47.6%. The average size of milk recorded herds is largest in Estonia at 74, in Latvia it is 13, and in Lithuania it is 17. Due to a reduction in the number of small farms (1-5 cows), the proportion of big farms is increasing gradually. Small keepers in Baltic countries go out of the dairy business because they get support from EU to do so. Milk production increases each year in all three countries. In Estonia, the average yield of milk across the various breeds in 2007 was 7052 kg with 4.15% fat (293 kg) and 3.36% protein (237 kg). In Latvia, the average yield of milk was 5478 kg with 4.37% fat (239 kg) and 3.37% protein (185 kg), while in Lithuania, milk yield was 5863 kg with 4.28% fat (251 kg) and 3.36% protein (197 kg). All three countries are trying to improve genetic merit. They are members of ICAR, while Estonia and Latvia are also members of INTERBULL. Estonia is the most liberal Baltic country in the area of cattle husbandry. It has achieved the best results in this sector, because government institutions have the least influence on the work and activities of cattle breeders.

Keywords: Baltic countries, cattle, cows, fat, milk yield, protein

Introduction

Historically as well as traditionally, the dairy sector has been, and remains, one of the most important agricultural activities in all three Baltic countries. Dairy cattle selection started to develop at the end of the 19th century and the first cattle breeding associations were established at the beginning of the 20th century and continued to be active until collectivisation. Even during Soviet times, all three countries maintained and developed dairy traditions, although most of the cattle were part of government owned collective farms and private owners were allowed to have one to three cows for personal needs.

After the Baltic countries declared independence in 1991, they started to establish family farms and agricultural cooperative communities, which initiated the creation of cattle breeder associations. These associations were established in the Baltic countries in 1992 and 1993. The new associations encountered a lot of issues in respect to cattle farm restructuring. The assistance received from foreign experts was greatly appreciated during the transition period. Foreign experts helped to establish cattle marketing systems and to strengthen breeding associations, so they could at least partially take over the functions carried out by the governments during the transition period. In Estonia, most of the breeding functions were transferred to the Estonian Cattle Breeding Association based on EU legislation. Meanwhile, in Latvia and Lithuania governments were actively regulating breeding functions and all required document forms and requirements. In order to use bulls effectively and conduct milk recording in their countries, Baltic countries joined ICAR. All three countries were granted a special ICAR stamp:

• Estonian Animal Record center was granted in 2006;
• Latvian public home animals breed and recording center was granted in 2005; and
• Lithuanian Ministry of Agriculture was granted in 2007.

The membership of the INTERBULL organisation is also very important to all three countries. Farmers can choose and compare bulls by using INTERBULL evaluation data. As of now only Estonia and Latvia are members of INTERBULL.

The main dairy organisations in the Baltic countries are:

• in Estonia there is one - Animal Breeders' Association of Estonia;
• in Latvia there are two – (1) Latvian Cattle Breeding Associations, and (2) Latvian Association of the Holstein Cattle;
• in Lithuania there are three – (1) Lithuanian Cattle Breeders Association, (2) Lithuanian Black and White Cattle Breeders Association, and (3) Lithuanian Red Cattle Breeders Association. They all are united by the umbrella organisation Chamber of Agriculture of the Republic of Lithuania.

The current population of Lithuania is 3.5 million, that of Latvia is 2.4 million and that of Estonia is 1.4 million. The most milk per capital (574 kg) is produced in Lithuania, and the least (370 kg) is produced in Latvia (Latvijas Statistika, 2008; SLRV, 2008). The quantity of milk produced per capita has been growing since 2000; in Lithuania it increased by 15.88%, in Latvia it increased by 5.94% and in Estonia it increased by 10.85%. In milk production, Lithuania exceeds Estonia by 10.1% and Latvia by 35.54% (Figure 1). The Lithuanian population consumes only 300 kg of milk and dairy products per capita and therefore the country should actively search for the markets for dairy products (SLRV, 2008).

Farmers realised that becoming part of EU provides bigger opportunities for dairy cattle farms but at the same time they also have to face higher competition. Using the EU supports for different programs, the Baltic countries have started active cattle farm modernisation, paying significant attention to cattle breeding. The herds of the growing farms continue to increase, but at the same time, small farmers availing of early retirement opportunities, are leaving the dairy business. Due to these changes, the number of dairy cattle started to decline in the Baltic countries. The average size of dairy herds is four cows in Lithuania, three cows in Latvia, and fifteen in Estonia. Compared to 2000, the overall dairy cattle number decreased, in Estonia by 20.5%, in Lithuania by 19.3%, and the lowest decrease is noted in Latvia at only 9.6% (Figure 2).

The decrease in cattle numbers can be explained by the decrease in the number of farmers. A lot of small farmers retired from the dairy business in Latvia resulting in a decrease of 44.3%, in Estonia the decrease was 34.7%, and in Lithuania it was 35.2% (Figure 3). It is expected that the number

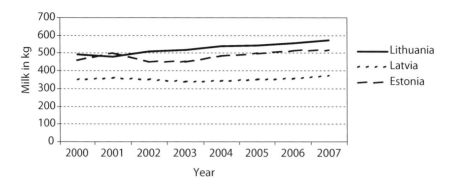

Figure 1. Milk production per capital in the Baltic countries.

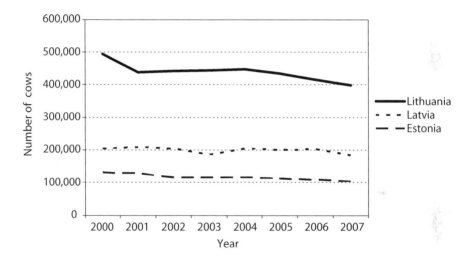

Figure 2. Dairy cattle numbers in Baltic countries.

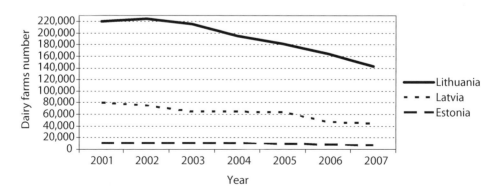

Figure 3. Dairy farm numbers in Baltic countries.

will continue to decrease in Lithuania next year (Eesti Statistica, 2008; Latvijas Statisika, 2008; SLRV, 2008).

Despite the decrease in the number of cattle, productivity is increasing in all three countries. The farmers realised that under free market conditions they have to increase the output of milk, improve the quality and invest in economically useful cattle. Therefore, demand for world class genetics has developed. Breeding heifers and cows, proven bulls, embryos and semen from Germany, Denmark, Finland, the USA and other countries are being imported into the Baltic countries. Until now most of the semen has been imported from EU countries, with the least amount from third world countries. The semen of bulls that are evaluated in the Baltic countries is not sold between the 3 countries.

The quality of the milk also improved due to better quality feed, better herd management and the sense of ownership. By using the breeding bulls that pass all the qualifications, an improvement was noticed in cow appearance, legs and longevity. Holstein, which was chosen for breed improvement, provided a high economic impact in all three countries. The biggest increase since 2000 in milk quantity was noted in Estonia, with an increase of 28.1%, in Latvia it was 13.7% and in Lithuania it was 17.4% (Figure 4) (Estonian Animal Recording Centre, 2008; VALDC, 2008; Žemės, 2008).

Since establishment of independence, all the Baltic countries have become concerned about the security and maintenance of the cattle breed gene pool. All three countries have a few breeds that they intend to preserve. In Lithuania the following rare breeds are being preserved: Lithuanian Native Whiteback, Lithuanian Light Grey, Lithuanian Black and White old genotype, Lithuanian Red old genotype. In Latvia, the Latvian Blue Cattle are being preserved and in Estonia the Estonian Native cattle are being preserved. A national coordinator is assigned to each country for national genetic resources preservation and coordinators maintain relationships with FAO. Rare cattle breeds are included into the World Agricultural Animals Variety Catalog (World Watch List; Scherf, 2000).

In regulated herds, the Holstein is the most numerous breed amounting to about 240,000 and continues to increase (Figures 5 through 7). In Figure 7, only 2.3% of the cows are shown to be Holstein in Lithuania; however this low number is due to the fact that Lithuanian Black and White, which makes about 70% of the cow population, is mixed with Holstein and has over 70% of Holstein breed mix. Hopefully, in 2009 through the cooperation of associations and government institutions, the name of the Lithuanian Black and White will be changed to Lithuanian Holstein. Lithuania has the biggest population of Red cows (about 62,000 cows).

The basis for breeding is milk recording. Based on the number of dairy animals, the most cows in milk recording are in Estonia at 90.9%, followed by Latvia at 69.4%, and Lithuania at 47.6% (Figure 8) (Estonian Animal Recording Centre, 2008; VALDC, 2008; Žemės, 2008).

The decrease in milk recorded herds was very extreme in Latvia and Lithuania. Before 1990, in Lithuania more than 500,000, and in Latvia more than 400,000, cows were recorded compared to 188,400 and 121,400 cows, respectively in 2007 (Latvijas Statistika, 2008; Žemės, 2008). In

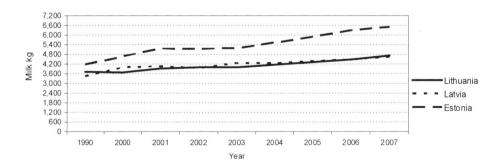

Figure 4. Milk yield per cow in Baltic countries.

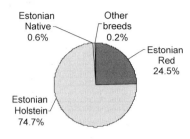

Figure 5. Cow population distribution in Estonia.

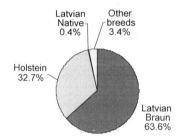

Figure 6. Cow population distribution in Latvia.

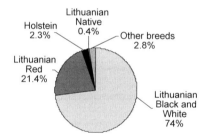

Figure 7. Cow population distribution in Lithuania.

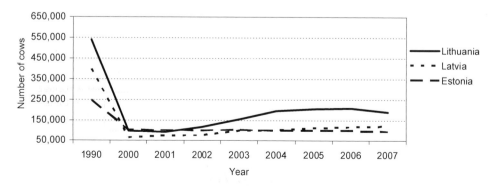

Figure 8. Numbers of cows in milk recording in Baltic countries.

Lithuania, associations and breeding enterprises are putting additional emphasis on milk recording; however, the lack of a common breeding strategy based on collaboration between different institutions led to the decrease of the number of recorded cows. The decrease in milk recorded cows was also impacted by the low milk price.

Along with the increase in dairy herd productivity, the milk production level of recorded cows is also increasing in all three countries. The production level in the milk recorded herds is increasing due to a consistent emphasis on breeding work performed by the associations. Since 2000, cow productivity increased most in Estonia (29.7%), with an average productivity in 2007 of 7,052 kg milk. In Latvia, the increase was 19.5%, with an average productivity in 2007 of 5,478 kg milk. The lowest increase in productivity was noted in Lithuania (17.3%), with an average productivity in 2007 of 5,863 kg milk (Figure 9) (Estonian Animal Recording Centre, 2008; VALDC, 2008; Žemės, 2008). The protein yield is increasing along with the milk yield. Since 2000, average protein content in milk increased most in Estonia (40.3%) with an average milk protein of 3.36% in 2007. In Lithuania, the increase was 26.4% with the same protein content as in Estonia (3.36%), and in Latvia, the increase was 21.1% with an average milk protein of 3.37% in 2007 (Figure 10) (Estonian Animal Recording Centre, 2008; VALDC, 2008; Žemės, 2008). Currently, the farmers' main focus is the increase of protein content in milk.

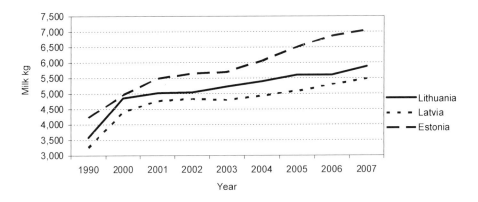

Figure 9. Average milk production of cows in milk recording in Baltic countries.

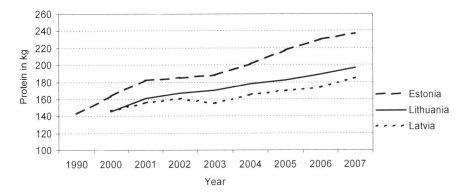

Figure 10. Average protein yield of cows in milk recording in Baltic countries.

The milk fat increase is also related to the overall milk yield increase. Since 2000, fat content in milk increased most in Estonia (27.6%) with a mean fat content of 4.15% in 2007. In Lithuania, fat content increased by 22.7% with a mean fat content of 4.28% in 2007, and in Latvia, fat yield increased by 18.8% with mean fat content of 4.37% in 2007 (Figure 11) (Estonian Animal Recording Centre, 2008; VALDC, 2008; Žemės, 2008).

Milk test laboratories are established in all three countries and are connected to the ICAR reference laboratories network. Laboratories are equipped with modern milk analysis equipment and are accredited according to ISO/IEC 17025 standards.

Before joining the EU, the Baltic countries raised the milk quote issue. All three countries negotiated acceptable milk quotas before joining EU. Milk quotas helped the three countries to control and administer the milk production and to balance milk volume. Lithuania received the biggest milk quota (Figure 12) (Productschap Zuivel, 2007). Due to milk and cheese consumption increases in Europe, in April 2008, the quotas were increased for all EU countries by 2% and it was agreed that the milk quota would be removed in 2015. At present, milk quotas in Lithuania are being auctioned. Since

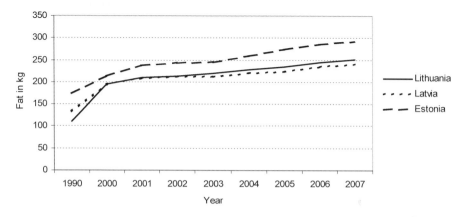

Figure 11. Average fat yield of cows in milk recording in Baltic countries.

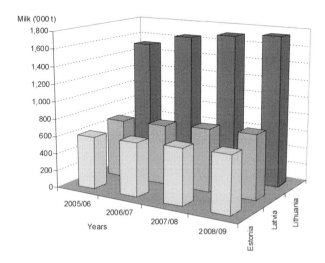

Figure 12. National milk quota in Baltic countries.

2007 it was decided to start sale by electronic auction. The farmers that have computers and internet access can acquire quota without leaving their houses. Milk quota in auctions can be purchased only by actual milk producers, which have usable land (for ten tons of milk quota they should have no less than 1 ha farm land) and have to be registered in the official register. The milk can be sold only to approved distributors. Quotas may be traded at an auction to be held three times a year. In Estonia, quota is traded with cows while in Latvia there is free trade in quota.

Of the three countries, Lithuania exports most dairy products, which consist of cheeses, milk powder and butter. Estonia and Latvia consume most of their dairy products internally and the main export dairy product in both countries is cheese. For all three countries, Russia is the biggest export partner. Most of the imports of dairy products is done by Latvia, while the quantities of imports by Lithuania and Estonia are similar.

In 2007, farmers were very happy with an increase in cow productivity and higher milk collection prices, hoping to increase investment in dairy products. Milk collection prices were increasing until the end of 2007, but in 2008, when the milk collection price significantly decreased, many farmers intended to retire from the dairy business. In the January – May period of 2008, the milk collection price decreased in Lithuania from €32.9 to €24.09 per 100 kg. In Latvia, it decreased from €33.83 to €27.72 per 100 kg. In all three Baltic countries, the average milk collection price is the lowest of all 25 EU countries. In May 2008, the milk collection price in Lithuania was the lowest with a difference of up to €11 per 100 kg compared to the EU 25 average price (Figure 13) (Lietuvos Respublikos Žemės Ūkio Ministerija, 2008a,b). The reason why the milk prices in Lithuania are the lowest is because after privatisation milk production landed in private companies. Those companies established milk collection prices that were not in favor of the farmers. The explanation the milk production companies provide for the low prices is that it is costly to collect the milk especially as most farms are small. Up till now, milk producers and farmers have not reached common ground. However, the solution might be in farmers' cooperation regarding milk collection and partial production.

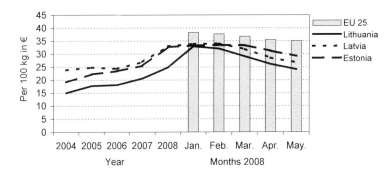

Figure 13. Milk collection prices in Baltic countries and average of 25 EU members.

Conclusions

The market is damaged and initiatives are slowed down due to tight government regulations and unwillingness to pass the responsibilities to associations. Such governing oriented structure encourages the use national money in a less efficient and economical way. Estonia is one of the most liberal countries within the Baltic countries with the least government interference in the breeding area and, therefore, shows the best results.

The sale of milk, milk products and breeding animals that presently takes place is handled separately within each of the three countries; however the countries should collaborate more in the breeding area. In the current globalisation process, Lithuania, Latvia and Estonia might consider some sort of cooperation to become more competitive in world markets.

Although the number of cows is decreasing in the Baltic countries, milk production is increasing, which calls for higher dairy product export. This is most applicable to Lithuania, which should be more active in export markets.

INTERBULL membership has positively impacted the Latvian and Estonian breeding sector. Lithuania must also become a member of INTERBULL to be able to perform a complete bull evaluation.

References

Eesti Statistica, 2008. Available at: http://www.stat.ee

Estonian Animal Recording Centre, 2008. Results of Animal Recording in Estonia 2007. Elmatar 2008. 52 pp.

Latvijas Statistika, 2008. Available at: http://www.csb.gov.lv

Lietuvos Respublikos Žemės Ūkio Ministerija, 2008a. VĮ Agro Rinka. Žemės ūkio informacijos ir kaimo verslo centras. Lietuvos žemės ūkio ir maisto produktų rinkos informacinė sistema. Nr. 8 (92) / 2008 m. 27 p.

Lietuvos Respublikos Žemės Ūkio Ministerija, 2008b. VĮ Agro Rinka. Žemės ūkio informacijos ir kaimo verslo centras. Lietuvos žemės ūkio ir maisto produktų rinkos informacinė sistema. Nr. 12 (96) / 2008 m. 30 p.

Productschap Zuivel, 2007. Statistisch Jaaroverzicht 2007. Productschap Zuivel, the Netherlands, 116 pp.

Scherf, B.D. (ed.), 2000. World Watch List for domestic animal diversity. 3rd edition. Food and Agriculture Organisation of the United Nations, Rome, Italy. Available at: ftp://ftp.fao.org/docrep/fao/009/x8750e/x8750e.pdf

Statistikos departamentas prie Lietuvos Respublikos Vyriausybės (SLRV), 2008. Available at: http://www.stat.gov.lt.

Valsts Aģentūra Lauksaimniecības Datu Centrs (VALDC), 2008. Available at: http://www.ldc.gov.lv

Žemės ūkio informacijos ir kaimo verslo centras (Žemės), 2007. Kontroliuojamų karvių bandų produktyvumas 2006-2007 metų (2006 10 01 – 2007 09 30) apyskaita 70. Vilnius 2007. 108 pp.

Cattle sector and dairy chain developments in Belarus

M. Ramanovich

IFCN Dairy Research Center, Schauenburgerstr. 116, 24118 Kiel, Germany;
mikhail.ramanovich@ifcndairy.org

Abstract

Milk production is one of the important branches of the Belarusian agricultural sector. In the beginning of the 90's, milk production in Belarus was negatively affected by the collapse of the socialist system. Both number and productivity of cows decreased significantly. Since 2001, improvement in the situation of the Belarusian dairy sector has been seen. Milk production increased due to growing productivity in large scale farms. In response to the growing milk production, Belarus could increase milk processing and the export of dairy products. The milk price paid to producers in Belarus is regulated by the government. As a consequence the price level in Belarus was significantly below the world market price for milk during the period of the analysis. For the analysis of farm economics, the methodology of the IFCN Dairy Research Center was used. Results showed that the cost of milk production in Belarus was low. Even with a milk price below the world market level, producers could realise a profit from milk production.

Keywords: Belarus, cattle sector, milk production, dairy chain, farm economics

Introduction

The dairy sector is one of the important branches of Belarusian agriculture. The aim of this study was to analyse the current situation and recent developments in the dairy sector in Belarus. The study contains an analysis of the whole dairy chain. Developments in milk production and processing, as well as developments of dairy trade and prices, were analysed. Additional attention was paid to the economics of milk production in Belarus in comparison with selected European countries.

Methods and data

In this analysis, the standard methodology of the IFCN Dairy Research Center was applied. The analysis was based on the official agricultural statistics from Belarus (http://belstat.gov.by), the FAO database (http://faostat.fao.org) and data collection by the IFCN Network (http://www.ifcnnetwork. org). For the farm level analysis, the concept of typical farms and the TIPI-CAL software (Hemme, 2000) were used.

Status and developments in Belarus dairy sector

Traditionally, the cattle sector was an important component of the agricultural sector of Belarus. In the Soviet Union, Belarusian agriculture specialised in milk and beef production for the common market. As a return service from other Soviet republics, the country received crop products for human consumption and feeding of animals. This specialisation ceased after the collapse of the Soviet Union, causing significant changes to the agricultural industry of Belarus (ZMP, 2002).

The structural changes to the dairy sector after the collapse of the socialist system in 1990 caused a significant reduction of dairy herds and cow productivity. As a consequence, milk production decreased dramatically. From 1990 to 2000, annual milk production decreased by 40% from 6.8 to 4.1 m t (Energy Corrected Milk). Since 2001, a consolidation of milk production has occurred (Figure 1). The volume produced increased significantly and amounted to 5.4 m t in 2007. This increase of milk production was achieved through an improvement of cow productivity while the number of cows was still decreasing. In 2007 the average milk yield per cow per year was 3.6 t while the total number of cows decreased to 1.5 million head.

In Belarus, a dual system of milk production exists. Producers can be divided broadly into two groups namely: large-scale farms and households. Large-scale farms are former collective farms (kolkhozes) and in most cases herd size is in the range of 400-800 cows. Households have milk production as a subsidiary business with 1-3 cows. The structure of milk production has changed over the years. In the beginning of the 90's, the share of big farms dropped from 75% to 60% and remained stable for several years. Since 2001 the importance of large-scale farms has increased again, and by 2007, they produced about 80% of the total milk supply.

Positive developments at farm level brought positive effects for the whole dairy chain. With the increased milk volume produced, the amount of milk delivered to dairies increased (Figure 2). In

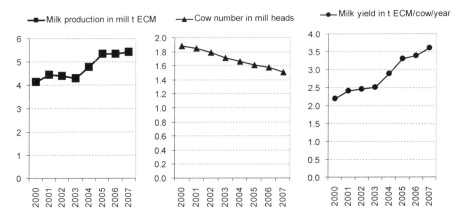

ECM= Energy Corrected Milk (4% fat, 3.3% protein)

Figure 1. Development of milk production in Belarus (Hemme, 2008).

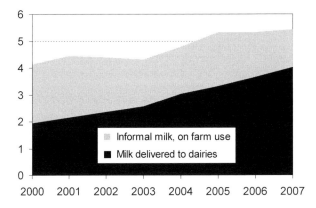

Figure 2. Evolution of milk delivery in Belarus (m t ECM) (Hemme, 2008).

2000, less than 2 m t of milk were processed to dairy products but by 2007, milk processing had doubled to 4 m t. The reason for this increase in milk delivered is an on-going industrialisation of milk production. Households producing milk, mainly for home consumption, are going out of business while at the same time large scale farms are continually increasing milk production.

With growing milk production and milk processing, the dairy sector in Belarus has become an important milk exporter on the world market. While milk imports for consumption has remained very low in recent years, milk exports have significantly increased (Figure 3). Between 1996 and 1999, Belarus was exporting dairy products equivalent to only 6 to 9% of the total milk produced. Since then the share of exports of total milk production has increased rapidly and in 2006 and 2007 about 35% of total milk produced was exported.

The current situation and developments in the dairy sector were strongly influenced by agricultural policies. The milk production sector, especially the large-scale farms, is governed by numerous state regulations. One of the most destructive regulations is the milk price policy as the raw milk price is fixed by the government. In addition, farmers cannot choose the dairy processor they prefer. In most cases, farmers are obliged to deliver their milk to the dairy processing plant in their administrative region. This interference of state institutions disturbs competition in the dairy market. Furthermore, the delivery obligations weakens the position of milk producers against milk processors.

The impact of price regulation is shown in Figure 4. The milk price paid to producers has been continuously below the world market price. The price difference in most years was about 5 US$

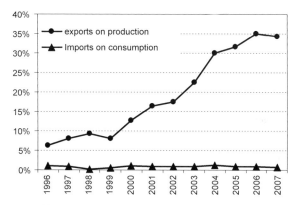

Figure 3. Dairy exports and imports (Hemme, 2008).

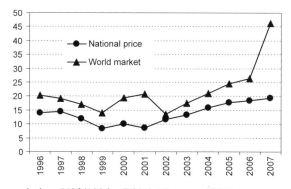

Figure 4. Milk price evolution (US$/100 kg ECM) (Hemme, 2008).

per 100 kg milk, but the gap increased as the recent years and growth in world market milk price was not fully transmitted in prices paid to milk producers in Belarus. As a consequence, the price difference in 2007 increased up to 25 US$/100 kg.

Dairy farm economics

To better understand the developments in the dairy sector of Belarus an international comparison of dairy farm economics was carried out. Belarusian milk producers were compared with producers from Germany, Poland, Bulgaria and Ukraine. The results are shown for 2006. The farms analysed differed in herd size, intensity of milk production and ownership. Milk production per cow per year is shown in Figure 5 together with the number of cows per farm.

The 3 typical farms analysed in Belarus were one household plot with 2 cows (BY-2) and two large-scale agricultural enterprises (BY-650 and BY-650++) each with a 650 cows but with different farm quality and herd management standards. The first of these (BY-650), had average management while the second (BY-650++) was managed according to best practice. The management quality affected the milk yield. The best-managed farm had an annual milk yield of about 5.6 t, while the farm with average management only yielded 3.3 t.

Farms in Ukraine were a household plot with 2 cows (UA-2) and one large scale farm with 641 cows (UA-641). The UA-2 had a milk yield per cow comparable to BY-2 and BY-650. The milk yield per cow of UA-641 was slightly below the level in BY-650++.

The two farms in Bulgaria were family farms. The 2-cow farm represents a household plot producing milk as a side business. This farm will probably go out of business in the next few years. The BG-34 was a larger family farm specialising in milk production and intending to increase herd size in future years. With 4.5 t per cow per year, the milk yield of the BG-2 was significantly higher than in the farms of the same size in Belarus und Ukraine. The milk yield per cow in the BG-34 was on the same level with the larger farm in Ukraine.

The farms in Poland were typical family farms with 15 (PL-15) and 65 cows (PL-65). The milk yield of PL-15 was 6.7 t per cow. The bigger farms had more intensive milk production with an annual yield of 7.2 t per cow.

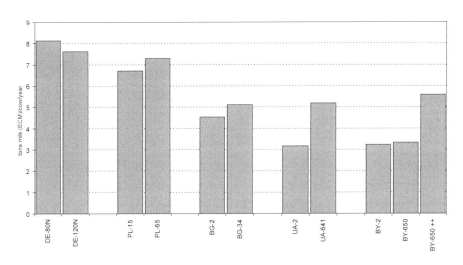

Figure 5. Milk yield on typical dairy farms (the number following the country designation is the number of cows per farm) (Hemme, 2007). The data refer to 2006.

The typical farms from Germany were average sized (DE-82) and somewhat larger (DE-120) family farms from Northern Germany. With more than 8.1 t milk per cow DE-82 had the highest milk yield of all the analysed countries.

To compare the economics of milk production in the selected countries, the costs of milk production and milk prices paid to producers were calculated (Figure 6). Costs of milk production consist of three cost elements: (a) costs from profit and loss account (comprising of all cash costs and depreciation costs for farm investments) – non-milk returns (comprising of cattle returns, beef returns, sales of manure, etc.); (b) opportunity costs for family-owned resources (land, labour and capital) used in the dairy enterprise; and (c) quota costs (comprising of rents and opportunity costs for owned quota). This was only relevant for countries with a milk quota system (Germany and Poland).

Generally, two cost levels could be identified namely a high cost level of 34-42 US$/100 kg milk, and a low cost level of US$ 15-22 /100 kg milk. All three farm types analysed in Belarus belonged to low cost category. Furthermore, relatively low costs of milk production were found for milk producers in Ukraine and on the larger farm in Bulgaria (BG-34). The costs of milk production in Germany and Poland, and the small family farm in Bulgaria were significantly higher. In comparison to high cost countries, milk producers in Belarus had a cost advantage of about US$ 20 per 100 kg milk.

The analysis also showed significant differences in milk prices paid to the farmers in the countries studied. The lowest milk price, on average about US$ 17 per 100 kg milk, was paid to small scale farms in Ukraine and Belarus. Large scale farms in Belarus received US$ 20-21 per 100 kg milk. The farms in Bulgaria, and the large scale farm in Ukraine, received between US$ 25-30 per 100 kg milk. The highest milk price of US$ 35-36 per 100 kg milk was paid to farmers in Germany and Poland. Compared to the milk price paid in high-priced countries, milk producers in Belarus received about US$ 15 less per 100 kg milk.

With the production costs and milk price level in 2006, large scale farms in Belarus were able to achieve a business profit from milk production. The small scale farm in Belarus was able to cover all cash costs but was unable to generate a return fully covering opportunity cost for own labour, land and capital. A similar situation was found in Ukraine, but due to a higher milk price, the large scale farm was able to secure a significantly higher profit than the large scale farms in Belarus. Also

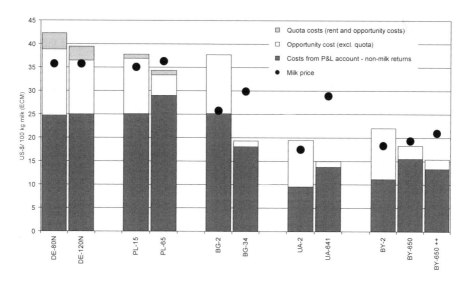

Figure 6. Economics of milk production in typical dairy farms (Hemme, 2007). The data refer to the year 2006.
Note: P & L = Profit and Loss.

in Bulgaria, the larger farm (BG-34) was much more profitable than the smaller farm (BG-2). Even with significantly higher milk prices, the profitability of milk production in Germany and Poland was the poorest. Among farms of the four countries analysed only one (PL-65) was able to the cover full economic costs of milk production.

Conclusions

The Belarusian dairy sector recovered after the collapse of the socialist system in the beginning of 90's and has achieved a significant increase in milk production in recent years. The main-stay of the dairy sector in Belarus are large scale farms producing most of the milk. With the increase of milk production, significant success in international dairy trade was achieved. In 2007, Belarus exported about 35% of milk produced in the form of dairy products and reached a self-sufficiency rate in milk of about 151%. Due to government regulation of the milk price, the price paid to producers was significantly below the world market level. Farm economic analysis showed that of milk production in Belarus was profitable. Large scale farms in Belarus were able to achieve a profit producing milk at low cost.

References

Hemme, T., 2000. IFCN – A concept for international analysis of the policy and technology impacts in agriculture. Ein Konzept zur international vergleichenden Analyse von Politik- und Technikfolgen in der Landwirtschaft. Landforschung Völkenrode, Sonderheft 215, Braunschweig.

Hemme, T. (ed.), 2007. IFCN 2007 Dairy Report 2007. International Farm Comparison Network, IFCN Dairy Research Center, Kiel.

Hemme, T. (ed.), 2008. IFCN 2007 Dairy Report 2008. International Farm Comparison Network, IFCN Dairy Research Center, Kiel.

ZMP, 2002. Landwirtschaft in GUS: Tier- und Pflanzenproduktion. ZMP Zentrale Markt- u. Preisberichtstelle, Auflage: 1.

Cattle sector and dairy chain developments in Ukraine

I. Ilienko

Association 'Ukrainian Agribusiness Club', Tbiliskiy Provulok 4/10, 03055 Kiev, Ukraine; ilienko@agribusiness.kiev.ua

Abstract

Due to the Ukraine's accession to the World Trade Organisation (WTO), there is need to expand into new markets and to increase the quality of meat and dairy products. In this paper, the major and most important tendencies, which may influence the competitiveness of the dairy chain in Ukraine are summarised. The rapid decrease in livestock numbers and the restructuring of the livestock production sector, which took place for the last 17 years, had objective reasons. While big farms were trying to get rid of loss-generating or low-profit activities, rural residents were maintaining relatively stable numbers of cattle, for the purpose of maintaining or improving living standards. As a result, the number of cattle is increasing within so called individual households. The individual households and small family farms are also an important supplier of meat and dairy products to the population and to the processing industry. The cattle population decreased in the two last years by 6% and 11%, respectively compared to corresponding period of the previous year. The most important trend in milk production by agricultural enterprises is the profitability of large farm enterprises, with on average more than 500 cows. This trend encourages companies, which already operate effectively, to further increase their scale of production. The greatest herd increases occurred in companies with more than 1000 cows. Across the whole dairy chain there are many weaknesses, starting with the prevalence of households' in the total raw milk supply and the low quality of this milk. There is also a pronounced seasonal pattern of raw milk production, a lack of investment in dairy farming, and underdeveloped logistics or infrastructure for milk collection, storage and distribution. This paper presents the official statistics, statements by sector players in the course of interviews and the results of recent empirical studies.

Keywords: cattle, dairy chain, Ukraine

Cattle populations and its spatial distribution in Ukraine by region

In the past decade a considerable decrease in the cattle population involving all breeds has taken place. This tendency is the result of the negative profitability of milk production and low prices for milk, low productivity and comparatively high primary costs of production.

In a regional context, most of the livestock and poultry population is concentrated in Vinnitsa, which holds a leading position in the livestock population as a whole and in the cow population in particular. The leading regions in livestock numbers as of January 1, 2008 are as follows (×1000 head): Vinnitsa 383.7, Khmelnitskiy 325.1, Lviv 317.5, Poltava 310.2 and Chernigiv 294.8 (Figure 1).

The greatest numbers of cows across the Ukrainian regions are found in agricultural enterprises in Central and North-Eastern Ukraine, with such regions as Poltava, Chernigiv, Kyiv, Cherkasy taking the lead. In the Central regions, North and North-East, livestock production is mostly specialised in growing cattle for dairy purposes and meat is a secondary product. In terms of livestock density (head per 100 ha), the Western regions take the lead (Figure 2).

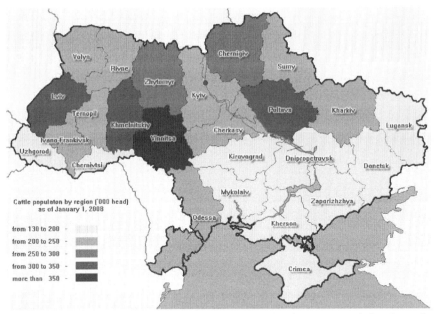

Figure 1. Cattle population in Ukraine by region (×1000 head). (State Statistics Committee of Ukraine, 2008)

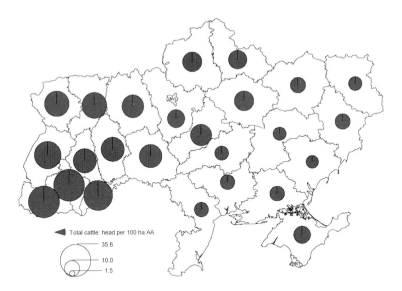

Figure 2. Spatial distribution of total cattle in Ukraine (head per 100 ha agriculture area)(Calculations of UCAB&vTi based on the data of the State Statistics Committee of Ukraine, 2008).

Genetic diversity

Altogether, 32 breeds are found in Ukraine. Of these, 17 are of milk and milk-meat types, namely Whitehead Ukrainian, Red Polish, Pinzgauer (all domestic breeds), Brown Carpathian, Red Steppe, Lebedin, Simmental (adapted breeds), Ukrainian Red-Piebald, Ukrainian Black-Piebald (breeds selected from foreign breeds); Ukrainian Red Milk (Ukrainian Fat-Milk type), North-East milk type of Brown breed (breeds created from foreign breeds); Golshtin/Holstein of European selection, Golshtin/Holstein of Canadian selection, Simmental of Austrian selection, Shvits breed, Angler and Airshire (all foreign breeds). The distribution of cattle breeds is summarised in Figure 3.

There are 7 domestic types of meat-breeds in Ukraine, which have been selected over time, namely Gray Ukrainian, Ukrainian, Volynska, Polisska, Simmental, Znamyanska, which account for 76% of all meat cattle. The remaining 24% are specialised foreign breeds: Aberdeen Angus, Hereford, Simmental (Australian and American selection), Limousin, Charolais, Blonde d' Àquitaine and Piemontese. Beef cattle breeds are mainly concentrated in Volyn (28%), Zhytomyr (14%), Rivne (9%), Kirovograd and Chernigiv (7%).

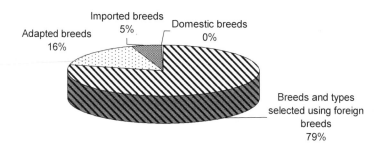

Figure 3. The structure of cattle breeds produced on large scale holdings (milk and dual purpose breeds) (Ministry of Agrarian Policy, 2008).

Structural changes in cattle numbers and milk production

As mentioned above, during last two years the cattle population decreased by 6%, and 11% respectively, compared to the previous year. An important trend is the changing proportions of cattle production between agricultural enterprises and private holders. The agricultural enterprises have reduced their share of the cattle population from 35% in 2000 to 22% in 2008 (Figure 4).

The large agricultural farms have been trying to cut losses caused by low-profit activities such as milk and beef production, while rural residents have retained a fairly stable number of cattle, in order to maintain or improve their standard of living. As a result, the majority of cattle is now kept by so called individual households. Individual households and small family farms are an important supplier of dairy and meat products to informal markets and to the processing industry. The considerable export potential of the dairy sector may only be fully realised after measurable improvements in productivity and in the quality of raw milk and dairy products are achieved. These conditions can best be met by large-scale milk production.

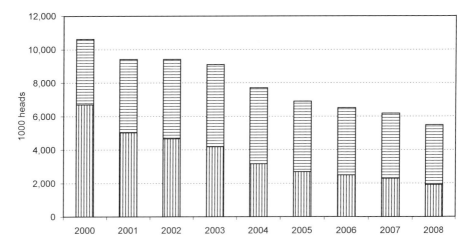

Figure 4. Numbers of cattle in Ukraine (State Statistics Committee of Ukraine, 2008).

Milk production by agro-enterprises

Structure of the cow population across farms

Improvements in profitability of milk production occurred mainly on large scale farms, which already operated effectively and they further increased their scale of production. The largest improvements were observed in companies, which already had more than 1000 cows. These large scale companies, specialising in dairy production, increased their number of cows (Table 1).

Dairy cow productivity

Agro-enterprises still need to increase productivity further. As Figure 5 shows, compared to western standards, the productivity of cows is low. On the other hand, as also shown in Figure 5, some dairy farms are capable of achieving a good level of productivity. It is important to note that average figures

Table 1. Structure of cow population across farms (data for 1 January each year) (based on the data of the State Statistics Committee of Ukraine, 2008).

	Number of agro-enterprises in the category			Change 2006 to 2008	Total number of cows by category (×1000 head)			Change 2006 to 2008
	2006	2007	2008	(%)	2006	2007	2008	(%)
up to 10	1,380	1,295	1,138	82	6.3	5.7	5.0	79
11-49	1,668	1,383	1,013	61	45.2	36.8	26.8	59
50-99	1,174	969	742	63	82.7	68.7	51.9	63
100-199	1,291	1,118	948	73	180	155.6	131	73
200-499	1,202	1,017	900	75	355.4	305.4	274.3	77
500-999	247	230	215	87	155.5	148.1	139.3	90
more than 1000	32	31	36	113	41.1	43.7	50.2	122
Total	6,994	6,043	4,992	-	866.2	764	678.6	-

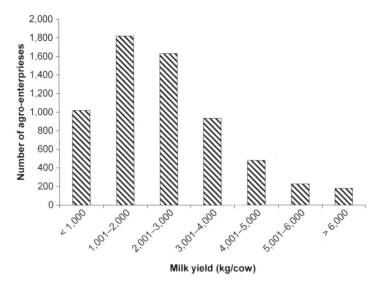

Figure 5. Milk yield distribution by agro-enterprises, 2007 (State Statistics Committee of Ukraine, 2008).

can be misleading in such a heterogeneous sector. While the average milk yield for dairy farms and households in 2007 was 3.7 t, the majority of agro-enterprises had yields no higher than 2 t. The average milk yield for agro-enterprises was 3.1 t while that for households was 3.8 t.

The number of enterprises with an annual herd milk yield above 6,000 litres is constantly growing. In 2006, the number of such companies was 181 and in 2007 it was 195 (Table 2). The share of such companies in the dairy sector is 3.6% of all holdings but their share of milk production is 13%. The general trends in agro-enterprises are a rapid decrease of cow numbers and a steady increase of milk yield per cow (Figure 6).

Table 2. Dairy cow productivity distribution among large and medium enterprises[1] (State Statistics Committee of Ukraine, 2008).

Productivity (kg /cow)	2006				2007			
	Number of agro-enterprises		Gross milk yield		Number of agro-enterprises		Gross milk yield	
	Number	Share in total (%)	×1000 t	Share in total (%)	Number	Share in total (%)	×1000 t	Share in total (%)
<1000	1,019	16.2	35.1	1.5	1,008	18.7	34.9	1.7
1,001–2,000	1,820	28.9	251.9	10.8	1,470	27.3	218.8	10.5
2,001–3,000	1,632	26.0	473.7	20.4	1,323	24.6	388.3	18.7
3,001–4,000	934	14.8	546.9	23.5	759	14.1	430.0	20.7
4,001–5,000	480	7.6	459.7	19.8	428	8.0	438.7	21.2
5,001–6,000	229	3.6	298.1	12.8	200	3.7	292.2	14.1
>6,000	181	2.9	257.9	11.2	195	3.6	271.2	13.1
Total	6,295	100.0	2,323.3	100.0	5,383	100.0	2,074.1	100.0

[1] >100 ha land or >50 employees.

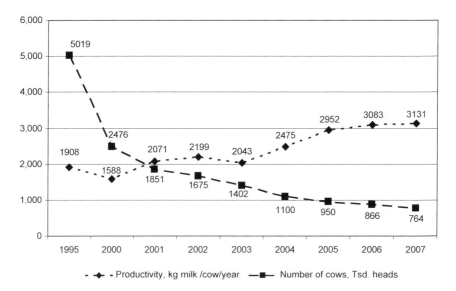

Figure 6. Dairy cow productivity and number of cows by agro-enterprises (State Statistics Committee of Ukraine, 2008).

Grazing system and/or confined

In Ukraine, year-round indoor management and partly indoor-grazing management are most common. In most holdings (98%), cows are kept in tie-stanchions and the rest are kept loose-housing systems. For beef cattle, both tie-stanchion and loose-housing methods are equally used.

Specialisation

Among agricultural enterprises, which have more than 50 employees and operate more than 100 ha farmland, specialisation in cattle breeding can be described as follows: on 40% of enterprises more than 60% of the total livestock kept are cattle, while on 10% of farms cattle account for 40% to 60% of all livestock kept. Pigs comprise the other sizeable share of farm livestock.

During the period 2005 to 2007, the number of agri-enterprises with more than 41% of the gross income derived from milk increased while the number with a lower milk revenue share decreased considerably (Table 3).

Slaughtering of cattle

In recent years, the number of cattle sold for slaughter has fallen considerably. As a proportion of total livestock product, cattle accounted for 47% in 2001 but decreased to 33% in 2007, and to 30% in 2008. It is evident that meat production from pigs and especially poultry, has an increasing market share. The meat supply originating from agricultural enterprises has also decreased, accounting for only 30% in 2007.

Young cattle, cattle of average fatness and cattle of high fatness (bulls over 400 kg and cows over 350 kg) are the common slaughter cattle and price categories. The average live-weight before slaughter of cows is about 170-180 kg depending on the season of slaughter, it is higher in autumn. In Ukraine

Table 3. Milk production specialisation on large and medium enterprises[1] (calculations based on the data of the State Statistics Committee of Ukraine, 2008).

Share of milk in the total revenue (%)	2005		2007		Change 2005 to 2007 (%)
	Number of agro-enterprises in category	Share in the total (%)	Number of agro-enterprises in category	Share in the total (%)	
1-10	2,166	46	1,339	41	62
11-20	1,327	28	758	23	57
21-30	792	17	524	16	66
31-40	424	9	339	10	80
41-50	162	3	171	5	106
51-60	75	2	89	3	119
61-70	23	0.5	34	1	148
71-80	6	0.1	9	0.3	150
81-90	2	0.0	5	0.2	250
>90	0	0.0	3	0.1	-
Total	4,677	100	3,271	100	70

[1] >100 ha land or >50 employees

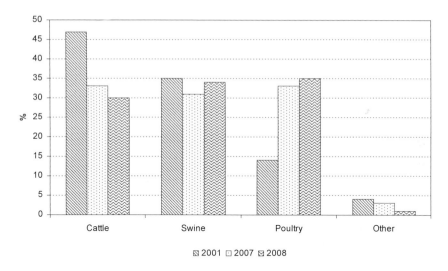

◪ 2001 ☐ 2007 ◪ 2008

Figure 7. Share of cattle, swine and poultry in the total stock sold for slaughter (live weight) (State Statistics Committee of Ukraine, 2008 – UCAB's estimation).

there are hardly any quality payment differences between breeds at slaughter, a consequence of the poor development of meat producing cattle breeds and production systems.

Dairy supply chain in Ukraine, its weaknesses and possibilities

Milk producing farms are mostly concentrated in the North-Central-West part of Ukraine and dairy plants to follow the same pattern. According to the official statistics, in 2007 Ukraine had about 600 dairy plants, most of them private, and about 3,700 milk producing agricultural large and medium enterprises (>100 ha land or >50 employees). However, these farms produced less than 20% of the total national raw milk (Figure 8).

Efficiency of ago-enterprises

The dairy chain in Ukraine has some distinct weaknesses, which start with the low efficiency and quality of milk production in the agro-enterprises. To evaluate the efficiency of agro-enterprises we used the Data Envelopment Analysis (DEA) method and made calculations with a program DEAP 2.1 (created by Tim Coelli).
A total of 870 enterprises were selected in which the share of milk in the total farm income was more than 30%. The average technical efficiency of these enterprises was 36.6% (of the possible maximum for the resources deployed), showing the huge potential for efficiency increases that exist at production level with a requirement for only minor additional resources. The efficiency of cattle breeding was 21% and 28% in 2005 and 2006, respectively.

Feed production for dairy sector

Feed production for dairy cows on Ukrainian farms generally takes place on a very extensive basis, which results in low feed quality and high total feed costs. Feed and forages account for about 50%-75% of the costs of cattle production. The decline in the cattle population since 1990 has led to a reduction in the domestic demand for forages for both agricultural enterprises and householders. As a result, there were some changes in land use, namely a reduction of natural meadows and cultivated pastures. There was also a reduction in the production of forage tubers and melons, sugar beet for forage, and maize silage. As a result of these changes the forage crop area decreased from 4 m ha to 3.1 m ha since 1990.

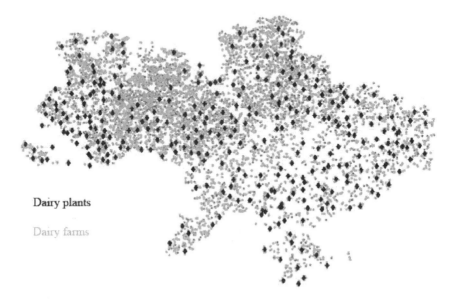

Figure 8. Dairy sector map of Ukraine (Nivievskyi *et al.*, 2007).

Production of hay, forage tubers, melon and sugar beet for cattle feeding is generally concentrated in private farms, while maize silage and hay production is mainly practiced in agricultural enterprises. Ukraine has about 2.4 m ha of hay and 5.5 m ha of pastures (State Committee of Ukraine for Land Resources, 2007), but yields are relatively low at 1.3 t/ha hay and 4.7 t/ha green mass. This area gives 2.1 m t of hay or 1.1 m t feed units[3]. Potatoes and vegetables are also used for feed. The total volume of such feed is about 2.5-2.7 m t feed units. Most of the fodder is produced on-farm. Feed additives are mostly imported. Ukraine also imports protein rich agro-industrial by-products and meals such as meat and bone meal, fish meal and soybean meal. These have better protein quality than sunflower meal.

Ukraine has great potential to expand domestic feed production. It is already exporting feed grains, sunflower meal and siftings but prices obtained for these products are low in comparison with world market prices.

The total supply of feed resources in Ukraine is estimated at 39-40 m t feed units, including 23-24 m t feed units of concentrate fodder. About 38% (ca. 14 m t feed units) of the total fodder stock is used for milk production. Concentrates and compound feeds are consumed mostly by poultry and pork (about 74%).

Production of mixed concentrates for cattle, as for other kinds of domestic animals, is increasing, but its share of total concentrate production is decreasing, mainly due to the considerable expansion of the poultry sector. In 2005, the concentrate feeds for cattle consisted about 599,000 t (19%), and in 2006 it was about 650,000 t (17%). The production was concentrated on the enterprises in the Sumy, Donetsk, Kharkiv, Poltava and Rivne regions, which produced more than half of the concentrate feeds for cattle.

Extension service and access to market information

An important agency in the development of the dairy sector is the extension service. There are only a few extension service centers in Ukraine. For example, IFC has a dairy project in Vinnytsia. Dairies and other agro-business operators are providing for the growing demand for specialised extension services. Private extension has also been developing in Ukraine. However, public extension has shown very little progress, despite the funding of such activities. On the national level, there is no official supply and demand statistics available for extension services (Nivievskyi and Strubenhoof, 2007). In most European countries, the extension services not only provide technical information, but also assist in business-planning, and provide consultancy on the preparation of official documents for budgets and subsidy applications. The development of extension services in Ukraine is controlled by legislation (Law of Ukraine 'On agricultural extension services' No 1807-IV on 17.06.2004). Such services may be financed from the public budget as well as donor projects. The current level of financing of the extension services in Ukraine is quite low. Besides, Ukrainian services often provide and make consultations only on general issues, mostly in social and private small business initiatives.

Pedigree Cattle

The current supply of pedigree cattle from the national selection schemes is far below the demand from dairy production enterprises. There are several ways of sourcing high quality pedigree cows (with productivity of 8-12 t milk/year):

---

[3] 1 feed unit = 1 kg of oat, 5 kg of green mass, 1-0.8 kg of grain; 500 kg cow needs at least 5 feed units per day. Usually to get 1 kg of milk a cow needs 0.5 feed units of fodder above maintenance. Thus, for 20 kg of milk a cow requires 15 feed units on average.

- Importation from abroad: three problems exist (a) import procedures, (b) must deal in large consignments, suitable only for large farms, (c) it is a likely that farmers will not obtain maximum yields because of poor practices and insufficient knowledge. The more productive a cow is, the more care it requires. In 2007, imports of cattle increased by 64% to 3,638 head compared to the previous year. The origins of imported pedigree cattle were: Germany 60%, Hungary 39%, the rest were from Denmark. The imported cattle from Hungary were 6% cheaper than cattle imported from Germany. The average price per head of pedigree cattle in 2007 was US$ 2,585.
- From domestic breeding organisations: the productivity of these is comparatively low. There are about 90 pedigree dairy breeding cattle farms in Ukraine. Most are in Kyiv and Cherkasy regions – 17 and 12 farms, respectively.
- Insemination of domestic cows with imported semen from pedigree bulls: the productivity of the offspring obtained would be lower than that of imported cows (ca. 9 vs. 12 t milk/year). However the cost would also be lower.

Logistic and infrastructure

Another important feature of the Ukrainian dairy supply chain is that logistics and infrastructure (milk collection, storage and distribution) are underdeveloped and expensive. Milk collection in most of cases, is done by dairies but sometimes local authorities or cooperatives may be responsible for collection. Usually, dairies use old trucks with 5 t milk tanks made from aluminum, in which it is difficult to maintain milk quality. In general the milk collection system in Ukraine is heterogeneous, as a large share of the raw milk is bought from households.

Raw milk quality

The prevalence of households in the total raw milk supply and the low quality of raw milk adds costs to the production chain, making it less competitive in the world arena (Niviewskyi *et al.*, 2007). Moreover, the Ukrainian quality standards are far from Western standards (Table 4), which limits the export possibilities for the dairies mostly to the former Soviet republics.
Milk from households usually classifies as milk of 2nd grade according to the Ukrainian standard. In the EU and USA such milk is not used for food production. At the same time, about 80% of all milk directed for further processing is supplied from households. The situation looks better in dairy farms, which deliver mostly 1st and higher grade milk (Table 5). Thus, a declaratively strong standards system doesn't simulate milk quality improvement in Ukraine.
The major weaknesses which determine efficiency of the dairy supply chain are: lack of investments in dairy farming mainly due to taxation of inputs (seeds, agrochemicals, machinery, etc.) via tariff and non-tariff import barriers, excessive regulation (e.g. certification), a lack of a market for farm land, a lack of marketing information and infrastructure, and an acute shortage of human capital. The pronounced seasonal pattern of raw milk production by households and dairy farms, and very low productivity of cows per lactation, adds problems and costs to dairy processors' operations.

Table 4. Raw milk quality standards for food production in Ukraine and EU (EC, 1992; US FDA, 2003).

	EU	Ukraine			
		Extra grade	Higher grade	1st grade	2nd grade
Plate count 30 °C (×1000 per ml)	≤100	≤100	≤300	≤500	≤3,000
Somatic cell count (×1000 per ml)	≤400	≤400	≤400	≤600	≤800

Table 5. Quality of milk sold by agro-enterprises to processors in 1st half of 2008 (DSTU 3662-97) (State Statistics Committee of Ukraine, 2008).

Quality	Milk sold × 1000 ton	Share in total (%)
Extra grade	0.5	0.1
Higher grade	236.9	27.3
1st grade	554.1	63.8
2nd grade	63.4	7.3
Offal (non varietal)	13.3	1.5

Production and consumption of dairy products and beef in Ukraine

Improvements of all the dairy supply and beef production chains in Ukraine has become urgent as there is scope to expand production and there is underused processing capacity. The distribution of dairy products for Ukraine is shown in Figure 9. The per-capita consumption of, and the export opportunities for, dairy products and beef are currently comparatively low but are likely to rise in the near future. Milk and milk products consumption per capita has grown since 2000 (Table 6). The small decrease in 2007 was caused by a considerable price rise for milk products.

Dairy products consumption in 2007 was 11% greater than in 2000, with regional variations of 2% to 44%. The highest consumption growth was observed in the Central, Eastern and Southern regions. That correlates with the higher income in these regions. In spite of the considerable increase in consumption of milk products, the level is still much lower than the biological norm of ca. 300 kg per year (the level of consumption defined as optimum for balanced nutrition), and is lower than in several European countries.

Whole milk is the most popular of all dairy products consumed. In monetary terms about 40% of the value of all dairy products consumed is whole milk, about 30% is cheese, about 15% is butter and spreads, and the rest is divided between ice-cream, canned milk and milk powder (11%, 4% and 2%, respectively) (Figure 10).

With the growing purchasing capacity of the Ukrainian population, and the availability of favourable opportunities for export, there is sizable potential for expanding production. With regard to beef, both production and consumption have decreased significantly, and the sale of cattle for slaughter has also decreased. Compared to 1990, production of beef in 2007 had decreased by 70% mainly because of reduced production by agricultural enterprises, even though householders doubled beef production in the same period. The share of beef in the value of all meat imports is only 5%. Before the Russian restrictions on Ukrainian dairy and meat products the major part of beef exports went to Russia.

Per capita consumption of beef in Ukraine was 10.8 and 10.2 kg in 2006 and 2007, respectively. Because of the shortage of beef for the meat processing industry, and reductions of import duties after WTO accession, an increase in imports of beef is expected.

Table 6. Production and consumption of milk and dairy products in Ukraine (State Statistics Committee of Ukraine, 2008).

	2000	2001	2002	2003	2004	2005	2006	2007	2000 to 2006 (%)
Production (mt)	12.7	13.4	14.1	13.7	13.7	13.7	13.3	12.3	97
Consumption (kg/year)	199	205	225	226	226	225	235	220	111

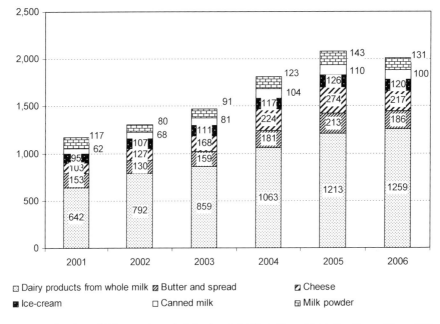

Figure 9. Production of dairy products in Ukraine (×1000 t) (State Statistics Committee of Ukraine, 2008).

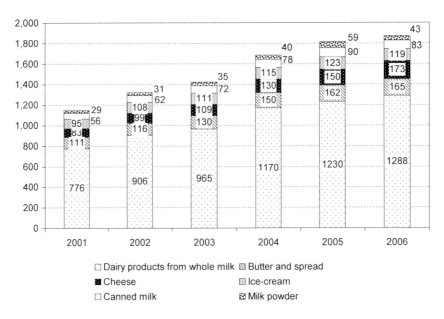

Figure 10. Consumption of dairy products in Ukraine (×1000 t) (State Statistics Committee of Ukraine).

Some aspects of dairy policies and producer support in Ukraine

Fixed agricultural tax

Agricultural producers may avail of special taxation benefits. According to the Ukrainian law 'On The Fixed Agricultural Tax', it is possible to change the tax order for agricultural producers from an ordinary set of taxes (various taxes including profit tax) to 'a fixed agricultural tax' (FAT) (Ukraine Government, 1998).

Agricultural enterprises are eligible to pay FAT if agricultural products account for over 75% of their revenues. This special tax regime replaces profit tax, land tax and some other taxes. The base for FAT is the normative value of a farm's agricultural land[4]. The *tax rates* are specified for three types of the agricultural land: (1) 0.15% of land value per year, for arable lands, meadows and pastures, (2) 0.09% of the value of perennial plantations, and (3) 0.45% of the value of water resources. FAT is paid monthly, but the payment rates vary so that 10% of the total payment is due in the first and second quarters, while 50% and 30% are due in the third and fourth quarters, respectively.

In addition, there is a special regime for the taxation of agricultural producers with value added tax. In 2008, the following main articles of the Law of Ukraine 'On The Value Added Tax' (VAT) remained in force (Ukraine Government, 1997). Their contents are summarised briefly as follows:

- Article 6.2.6: VAT equals zero for producers of meat (in live weight) and milk who sell their products directly to processing enterprises.
- Article 11.21: VAT amounts to be paid by meat and milk processors to the state budget are redirected to meat and milk producers as a subsidy proportional to the delivered raw materials.
- Article 11.29: VAT amounts from selling agricultural products (except meat in live weight, and milk) remain on the accounts of agro-producers and may be used for operational needs.

The FAT will stay in force until 2010. Meanwhile, the special order of VAT payments are prolonged for the year 2008.

Thus, the first and the biggest subsidy element for raw milk producers come from VAT collected from the dairy processing industry (already mentioned above). VAT received by dairy processing enterprises from selling dairy products is retained on a special account and paid to agricultural producers selling their unprocessed milk to processing enterprises.

An important aspect of these subsidies is that they do not comply with WTO requirements, since they are domestic support measures (WTO, 1994; Nivyevskyi *et al.*, 2008). The VAT regime for milk and meat producers will change with WTO accession (most likely beginning of 2009).

Special regime of Pension Fund payments for agricultural enterprises

FAT payers pay to the Pension Fund at special rates (19.38% in 2008) with 20% annual increase until the common rate will be reached (32.3%) (Ukraine Government, 2004). Then in a few years, FAT payers are expected to pay the full payment rate to the Pension Fund.

Subsidy for cattle grown and sold

Cattle of 390 kg minimum live weight for agro-enterprises and 330 kg for households are subsidised at 2.9 Ukrainian Hryvnia (UAH)/kg (€1 = 7.4 UAH; US$1 = 5.8 UAH).

---

[4] Land value is determined according to quality and potential productivity and, therefore, can vary substantially from farm to farm. The average land value in Ukraine for FAT purposes on the 1st January 2008 is 9,179 UAH/ha, ranging from a maximum of 12,708 UAH/ha in AR of Crimea to a minimum of 6,664 UAH/ha in Zhtomyr (excluding Kiev city, Sevastopil and Crimea).

Special subsidy for heifer livestock

A special subsidy for breeding heifers from suckler-cows, which are bought from households to increase livestock numbers is paid to agro-enterprises at a rate up to 5 UAH/kg.

Special subsidy for livestock

A special subsidy is paid for beef cattle if the meat-productivity of the animal is correctly identified. Producers of eco-tested milk delivered to dairy processing factories for producing baby food obtain a subsidy at the rate of 500 UAH/t milk.

Partial interest rate compensation

Agricultural enterprises may receive compensation for interest rates for short-term credit obtained in national or foreign currencies for covering production costs (e.g. purchases of fuel, feed, spare parts, fertilisers, pesticides, insurance payments, etc.), and for long-term credit obtained in national or foreign currencies for financing investments in fixed capital.

Partial compensation of agricultural machinery costs

Under this scheme, the Government compensates 30% of the price of domestically produced agricultural machinery to agricultural producers and enterprises of the food processing industry.

Partial compensation of insurance payments

Farms receive 50% compensation for insurance payments.
It is widely accepted that the Ukrainian problem is not the level of taxes but the complexity of tax management and tax control. To date, there is little transparency in the performance of the local tax administrations, especially regarding VAT compensation. In this way, tax authorities may prolong terms of exporters' examination before rewarding an expected compensation. Agricultural exporters are also very familiar with this practice.

Conclusions

The livestock and milk production industries have considerable potential to increase production capacity, expand markets and adopt technologies for production and efficiency improvements at all levels of the production chain. Demand for dairy and livestock production will further increase. Simultaneously, exports will grow. Following the prohibition of dairy and livestock products to Russia in 2006, more and more producers are now receiving individual export licenses after a series of inspections by representatives of the Russian authorities. In addition, livestock and dairy producers, together with governmental bodies, are working intensively to achieve access to European markets for their products. This requires further harmonisation of domestic and European quality standards with more distinct boundary setting at the legislative level by the central quality controlling bodies, quality improvements of dairy products, and improvement of storage and transportation systems for all stages of production.
For agricultural enterprises, there is a demand for improvements in production, optimisation of available resource use, increase in efficiency by introduction of new technologies, improvement of the skills of the employees of the farms, improvements in management, further restructuring of the industry, and finally an increase in foreign market share by individual enterprises. Areas near

dairy plants that have been modernised, have a higher productivity growth, mainly due to enhanced technological progress. Processors requiring a dependable supply of high quality raw milk assist farms to make the necessary investments for equipment, production resources and other needs. The main drawbacks to the state programs of industry support are instability in the financing priorities, the necessity for annual re-approval for fund distribution together with insufficient information and consultation with agricultural producers. Besides, for the stabilisation of livestock and milk production, there is a need for a stable tax system, transparency and uniform distribution of state support, and decreases in the time frames for investments returns.

References

European Commission (EC), 1992. Council Directive 92/46/EEC of 16 June 1992 laying down the health rules for the production and placing on the market of raw milk, heat-treated milk and milk-based products. Official Journal of the European Union L 268: 1-32.

Nivyevskyi, O. and Strubenhoff, H., 2006. Barriers to investment in the agriculture and food sector in Ukraine. IER Policy Paper. Available at: http://ierpc.org/ierpc/papers/agpp5_en.pdf.

Nivievskyi, O., Ilienko, I. and Ryzhkova, M., 2007. Dairy supply chain in Ukraine: bottlenecks and directions for development. Presented at the IAMO Forum, Halle (Saale), Germany.

Nivyevskyi, O., Von Cramon Taubadel, S. and Brümmer, B., 2008. Subsidies and technology change of Ukrainian dairy farms: spatial dependence in the components of productivity growth. Presented at the Vth North American Productivity Workshop 25-27 June 2008, New-York, USA.

State Statistics Committee of Ukraine, 2008. Website available at: http://www.ukrstat.gov.ua/.

Ukraine Government, 1997. Ukrainian law 'On The Value Added Tax' (VAT) – No. 168/97-BP as of April 03, 1997. Available at: www.rada.gov.ua.

Ukraine Government, 1998. Ukrainian law 'On The Fixed Agricultural Tax' (FAT) - No. 20-14 as of December 17, 1998. Available at: www.rada.gov.ua.

Ukraine Government, 2004. Ukrainian law 'On amendments to some laws on taxation of agricultural enterprises' – No. 2287-IV as of December 23, 2004. Available at: www.rada.gov.ua.

US FDA, 2003. Grade 'A' pasteurized milk ordinance 2003 revision. United States Food and Drug Agency, CFSAN, USA. Press releases; DSTU 3662-97. Available at: http://www.cfsan.fda.gov/~acrobat/pmo03.pdf

WTO, 1994. Uruguay Round Agreement on Agriculture. Available at: http://www.wto.org/english/docs_e/legal_e/14-ag.pdf

Zubets, M. and Melnychuk, D., 2004. Presentation on genetic resources of Ukrainian cattle. Ukrainian academy of agrarian sciences.

Cattle sector and dairy chain development in Slovakia

M. Stefanikova

Slovak Association of Dairy Farmers, Výstavná 4, 949 01 Nitra, Slovakia; szpm@agrokomplex.sk

Abstract

This paper presents the current situation and a perspective of the cattle sector and dairy chain development in Slovakia under the impact of the Common Agricultural Policy of the European Union. The first section is dedicated to dairy farming, including data on cattle numbers, production unit numbers, cattle breeds, milk yields, milk production, dairy farmer numbers, milk sales, milk quotas, milk prices and support policy. The second section deals with the milk processing/dairy sector including the number of dairies, production of milk and dairy products, and milk balance. The third section deals with issues of consumption and promotion of milk and dairy products. The fourth section presents some results on the economic efficiency of milk production in Slovakia and in European Dairy Farmers (EDF) countries. The fifth section discusses issues of the dairy sector in the EU including the 'Health Check of the CAP', soft-landing' and the Slovak position to the proposed measures.

Keywords: cattle breeds, cattle sector, dairy farmers, milk prices, dairy products consumption, Slovakia

Introduction

The Slovak cattle sector has undergone important developments in recent years. The following sections highlight the developments in (milk) production and prices, the milk processing sector, consumption and marketing of dairy products and economy.

Production and prices

Numbers of cattle and production units

Since 1989 the cattle population has decreased by 65%, from approximately 1.5 million to 0.5 million (Figure 1). The main reasons for this large decrease in cattle numbers were political, economic and social changes. The transition period and accession of Slovakia to the European Union were the most important milestones in this development. Big changes occurred in the cattle sector in terms of structure, ownership, producer and cattle numbers. Improvements in housing, feeding and milking technologies, but also in genetics and nutrition, positively influenced milk yields. On the other hand, dairy farmers had to invest heavily to comply with strict hygienic and environmental requirements. Big changes occured also in the production volume and in the overall relationships among those involved in the dairy chain.

Negative impacts on the cattle sector started with enormous increases of input prices at all production levels. Input prices led to inceases of consumer prices, which had a negative impact on consumption of milk and dairy products. At the same time, especially after the EU accession, the volume of imported dairy products increased. All these circumstances led to excessive over-production, which had a negative impact on farm-gate prices. Many dairy farmers, mainly those with low productivity, gradually ceased milk production.

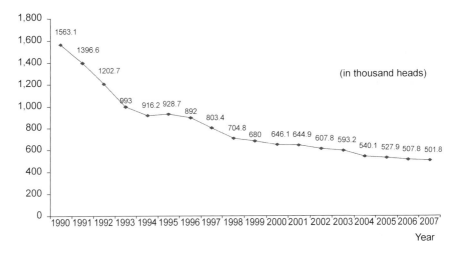

Figure 1. Cattle numbers (×1000) in 1990-2007 (Slovak Association of Dairy Farmers, 2008).

According to the figures of the State Breeding Institute (Table 1), there were 519,584 cattle in Slovakia at the end of 2007, managed in approximately 11,500 production units (including households producing milk for own consumption). Over 44% of all cattle are in herds of >500 head. The average herd size is 45 head/production unit. But if herd sizes below 10 are extended, then the average herd size is 258.

Cattle breeds

Fleckvieh (Simmental) and Pinzgauer were traditional Slovak breeds. Nowadays these breeds are characteristic of, and mainly confined to, mountainous areas. Fleckvieh cattle decreased by 29% and Pinzgau cattle decreased by 4%. Prevailing breeds for intensive milk production units in lowland conditions are Black or Red Holstein which comprise 56% of all cattle. Other milking breeds represent <18%. Meat breeds represent 8%, and 3% of the cattle have no breed classification (Table 2).

Table 1. Number of cattle and production units in 2007 (data provided by the State Breeding Institute).

Category (herd size)	Cattle production units		Cattle	
	Number	%	Number	%
1-10	9,518	82.6	19,474	3.8
11-20	326	2.8	4,688	0.9
21-50	248	2.2	8,187	1.6
51-100	207	1.8	15,640	3.01
101-200	327	2.8	48,308	9.3
201-500	588	5.1	193,083	37.2
over 500	313	2.7	230,204	44.3
Total	11,527	100.0	519,584	100.0

Table 2. Cattle breeds in Slovakia (data provided by the State Breeding Institute).

	Breeds						Total
	Fleckvieh (Simmental)	Pinzgau	Black and Red Holstein	Other milk breeds	Meat breeds	Other breeds	
Head	151,956	18,326	295,924	3,626	39,270	17,900	527,002
%	28.83	3.48	56.16	0.68	7.45	3.4	100

Beef breeds

There were 39,270 beef cattle at the end of 2007, representing 19 beef breeds. The most common breeds are Charolais 45%, Limousin 34%, Simmental 6%, Blonde d`Áquitaine 3% and Piemontese 3%.

Milking cows

The number of milking cows dropped from 542,800 in 1990 to 180,600 in 2007, a decrease of 67% in 19 years (Figure 2). That decrease is still continuing with a decline of 6% or 11,900 head in 2007. Due to the current economic and political situation in the EU dairy sector, a continued reduction in milking cows is expected in future. According to present developments it is estimated that there will be further drop of approximately 5%, or 9,000 milking cows in 2008.

Average milk yields

Last year annual milk yield reached 5,951 kg/cow. While this represents a continuing increase, it is nevertheless still rather low. In 2007, milk yield increased by 5% or 281 kg/cow. Continuous improvement in milk yield is expected in future due to improvements in husbandry technologies,

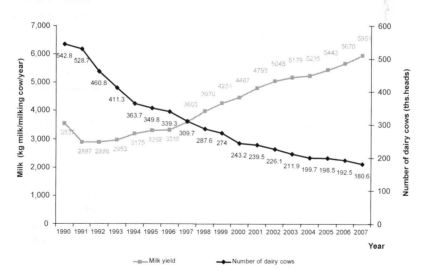

Figure 2. Number of milk cows and milk yields (1990-2007) (Slovak Association of Dairy Farmers, 2008)

housing, feeding and herd management. The estimate of that improvement is approximately 250 kg milk per cow (+ 4% p.a.) in future years.

Milk yields of cattle breeds

Holstein is the most common dairy breed and has the highest milk yield reaching 7884 kg of milk per cow in 2007 (Table 3). Slovak Simmental and Pinzgauer have a higher fat content and therefore are usually used in crossbreeding programmes.

Table 3. Milk yields of cattle breeds (Data provided by the State Breeding Institute).

Breed	Number (head)	Milk (kg)	Fat (%)	Protein (%)
Holstein	25,640	7,884	3.9	3.2
Holstein with crosses	75,931	7,198	4.0	3.2
Slovak Simmental	25,045	5,223	4.1	3.3
Pinzgaeru	1,072	4,323	4.0	3.3
Pinzgauer with crosses	4,553	4,534	4.0	3.3
Brown Swiss	147	6,414	4.0	3.3
All breeds	113,175	6,517	4.0	3.2

Suckler cows

There were 35,400 suckler cows at the end of 2007. Since 1997, the number has increased almost four fold from 10,100 (Figure 3).

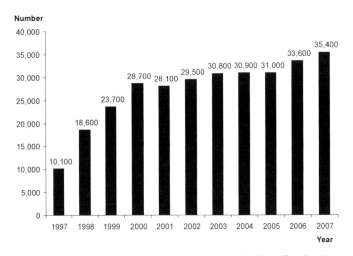

Figure 3. Number of suckler cows (1997-2007) (Slovak Association of Dairy Farmers, 2008).

Milk production

The large reduction in dairy cows had a profound impact on milk production. While in 1989 more than 2 m t of milk were produced, nowadays it is only half that (1,074 m t in 2007). During 2007 milk production decreased by 2% (Figure 4).

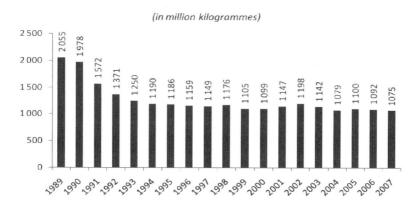

(in million kilogrammes)

Figure 4. Milk production in Slovakia (1989-2007) (Slovak Association of Dairy Farmers, 2008).

Number of milk producers/dairy farmers

At present there are 680 milk producers/dairy farmers in Slovakia (Figure 5). Due to unfavorable conditions in the dairy sector, mainly increases in of input prices, stagnation of milk prices, and absence of support, 146 dairy farmers (18%) stopped milk production during the last 4 years. If the present economic and political conditions continue, it is expected that the number of dairy farmers will continue to decrease.

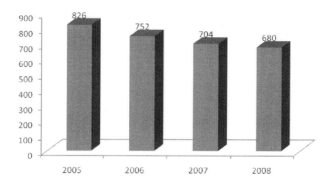

Figure 5. Number of milk producers/dairy farmers (2005-2008) (Slovak Association of Dairy Farmers, 2008).

Sale of milk

In 2007, Slovak milk producers/dairy farmers sold 974 million kg of raw cows milk (Figure 6). This includes both delivery to purchasers and direct sales. There have been only small differences in the amount of sales in recent years. The difference between the last two years was only 0.4% (4 m kg). In the last two years 95% of the milk sold was classified as Q – superior and I. class quality.

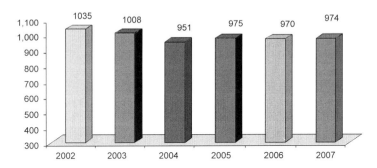

Figure 6. Sales of milk (2002-2007) (Slovak Association of Dairy Farmers, 2008).

Milk sales

The seasonal and monthly distribution of milk sales from 2004 to June 2008 is shown in Figure 7. Based on this analysis it is estimated that the volume of milk sold this year will be similar to last year.

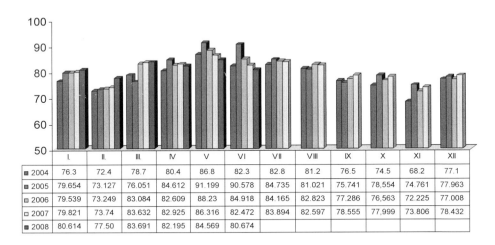

	I.	II.	III.	IV	V	VI	VII	VIII	IX	X	XI	XII
2004	76.3	72.4	78.7	80.4	86.8	82.3	82.8	81.2	76.5	74.5	68.2	77.1
2005	79.654	73.127	76.051	84.612	91.199	90.578	84.735	81.021	75.741	78.554	74.761	77.963
2006	79.539	73.249	83.084	82.609	88.23	84.918	84.165	82.823	77.286	76.563	72.225	77.008
2007	79.821	73.74	83.632	82.925	86.316	82.472	83.894	82.597	78.555	77.999	73.806	78.432
2008	80.614	77.50	83.691	82.195	84.569	80.674						

Figure 7. Delivered milk (2004-2008) (Slovak Association of Dairy Farmers, 2008).

Milk purchase

At present there are 48 dairy processing companies purchasing milk and registered by the Agricultural Payment Agency. Of these, 19 are production and trade cooperatives and 29 are dairies and business companies.

National Reference Milk Allocation - milk quota

As a result of the negotiation process with the EU, the initial milk quota was 1,013,316 m kg starting in 2003/2004 and continuing until 2005/2006. For the quota year 2007/2008, an additional 27 m kg were allocated as a so called 'restructuring reserve', increasing the national milk quota to 1,041,788 m kg. After the approval of the Council Regulation this was further increased by 2% to a national quota of 1,061,603 m kg. There are no regional quotas in Slovakia. The national milk quota is divided into a milk delivery quota (99%) and a quota for direct sale of milk (1%). The reference fat content is 3.71%.

Milk quota fulfillment - quota years 2004-2008

The proportion of the national milk quota filled in successive years was: 95% in 2004/5, 97% in 2005/6, 93% in 2006/7, and 95% in 2007/8. With regard to milk sales, it is estimated that the current milk quota will be filled to approximately 96% in the 2008 quota year. Therefore, Slovakia is opposed to the current proposal of the European Commission to issue additional milk quota.

Milk quota administration

The main stakeholders in quota administration are:
* Ministry of Agriculture SR as the State Authority;
* Agricultural Payment Agency as the Competent Executive Authority;
* 680 milk producers/dairy farmers;
* 48 milk purchasers.

Main features of milk quota

* milk quota is bound/tied to the company and to cows;
* milk quota is not subject to free trade (selling/buying), neither is it subject to rent;
* milk quota is not included into the company estate;
* milk quota is a subject of allocation from national reserve or quota transfer.

Main administration procedures of milk quota

* new milk quota allocation;
* additional milk quota allocation;
* milk quota transfer.

New milk quota allocation

* it is allocated to new dairy farmers from the national reserve based on application;
* maximum of new milk quota is 50,000 kg/applicant.

Additional milk quota allocation

Additional milk quota is allocated to existing dairy farmers from the national reserve based on application. Eligible applicants are dairy farmers who filled their existing milk quota to at least 95%. Applicants can apply for additional milk quota to a maximum of 15% based on the existing milk quota. (Payment Agency considers the application and allocates the applicants requests according to the available milk quota in national reserve equally/accordingly)

Milk quota transfer

Transfer of milk quota is possible only through purchase or renting of the whole farm or purchase of milking cows

Milk prices

Figure 8 shows the average milk price from 2004 until June 2008. Calculated using the monthly exchange rate between the Slovak crown and Euro, the milk price was approximately €33.25/100 kg in June. From January to June the price dropped by 15%. Slovakia is scheduled to join the Euro zone on 1st January, 2009. Due to the strengthening of the Slovak currency, there has been an unusual development in the price comparison and its calculation in Slovak crowns and in Euro. While the price in Slovak crowns decreased in recent months the price in Euro increased.

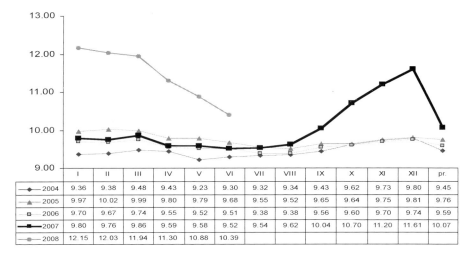

	I	II	III	IV	V	VI	VII	VIII	IX	X	XI	XII	pr.
2004	9.36	9.38	9.48	9.43	9.23	9.30	9.32	9.34	9.43	9.62	9.73	9.80	9.45
2005	9.97	10.02	9.99	9.80	9.79	9.68	9.55	9.52	9.65	9.64	9.75	9.81	9.76
2006	9.70	9.67	9.74	9.55	9.52	9.51	9.38	9.38	9.56	9.60	9.70	9.74	9.59
2007	9.80	9.76	9.86	9.59	9.58	9.52	9.54	9.62	10.04	10.70	11.20	11.61	10.07
2008	12.15	12.03	11.94	11.30	10.88	10.39							

Figure 8. Milk prices in Slovakia (2004-2008) (Slovak Association of Dairy Farmers).

Milk price comparison in EU member states

Figure 9 shows the average milk price for 2007 in the 26 individual EU member states. The average Slovak milk price was €28.97/100 kg. It was less than the milk prices in surrounding countries – especially Czech Republic, Austria, and Poland.

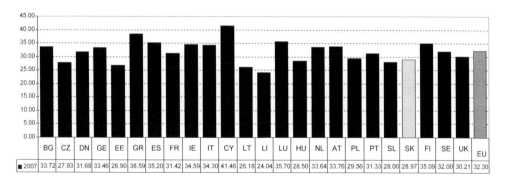

Figure 9. Milk prices in EU states (2007) (DG AGRI - Analysis Circa).

Support of the dairy sector in Slovakia

Subsidies or other support payments have not been provided to the cattle sector in Slovakia since 2004. It was one of many bad decisions of the previous government that resulted in loss in competitiveness of Slovak dairy farmers. Farm companies, specialised in dairying, were most affected, leading to an 18% drop of dairy farmers during the period.

The Slovak Association of Dairy Farmers always defended the interests of dairy farmers, but without any success with the previous government. After the new government was elected, agriculture was defined as one of the national priorities and support for animal production was introduced.

The Ministry of Agriculture introduced measures for the revitalisation of animal production and subsequently the 'Government Regulation on Support in Agriculture' by CNDP based on payments per Livestock Units was approved (Table 4).

Table 4. Government Regulation on Support in Agriculture by CNDP based on payment per Livestock Unit (Slovak Association of Dairy Farmers).

Category	Coefficient[1]	De-coupling date
Calves up to 6 months	0.2	De-coupling 31. March 2008
Cattle 6-24 months	0.6	De-coupling 31. March 2007
Bulls, oxen and heifers older then 24 months	1.0	De-coupling 31. March 2007
Suckler cows older then 24 months	1.0	Coupling 15. April of actual year
Sheep and goats older then 12 months	0.15	De-coupling 31. March 2007
		Coupling 15. April of actual year
Milk quota	0.1998	De-coupling 31. March 2007

[1] Coefficient for calculation of animal numbers for livestock unit numbers.

127

Milk processing sector

Number of processing dairies

There are 28 dairies processing more than 2 m kg milk, annually. Another 31 dairies process between 0.5 to 2 m kg, and further 61 dairies process less than 0.5 m kg milk, annually.

Production of milk and dairy products

According to the statistical data, dairy processing increased considerably in 2007 (Table 5). The individual product increases were: skimmed milk powder 44%, processed cheese 18%, butter 9%, fresh cheese 10%, liquid milk 6%, fermented milk products 6%, and whole milk powder 4%.

Table 5. Production of milk and dairy products (t) (Research Institute of Agricultural and Food Economy, 2008).

Year	Liquid milk	Natural cheese	Processed cheese	Cream	Fermented products	Butter and milk fat products	Total milk powder	Skimmed milk powder	Whole milk powder
2004	292,712	37,105	11,747	31,190	50,279	13,131	11,550	6,772	4,719
2005	246,873	43,447	10,630	34,535	52,263	10,034	12,856	5,801	6,634
2006	238,331	47,879	11,595	33,670	51,305	10,689	11,954	5,705	5,571
2007	252,279	44,669	13,641	34,619	54,180	11,690	14,285	8,207	5,808

Consumption and promotion

Consumption of milk and dairy products in Slovakia

Consumption of milk and dairy products in Slovakia has decreased by almost 100 kg per capita per year over the last 15 years, a drop of 39% (Figure 10). Whereas the average Slovak person consumed 253 kg of milk and dairy products in 1989, it was only approximately 153 kg in 2006. The decrease in milk consumption was caused mainly by the reduced purchasing power of the population and the wide range of soft and energy beverages on the market. In addition, the dairy sector and the previous governments did not invest in advertising to consumers. In terms of product range consumption, there was a decrease in liquid milk and an increase in cheese (Table 6).

Promotion of milk and dairy products

Despite the fact that milk producers and milk processors fight on different sides of the battle field, all fight in the same war. The arguments of the processors on high investment requirements and really strong pressure from the retail chains that are forcing down the milk price are accepted. It is also acknowledged that the appreciation of the Slovak currency has had a negative impact on dairies with export activities. Furthermore, it is acknowledged that the fight against the retail chains is like fighting against wind-mills. Efforts are underway to identify common areas of interest between producers and dairies which might improve the overall situation of the dairy sector in Slovakia. Thus, the Slovak Association of Dairy Farmers – as the representative of milk producers/dairy farmers, and the Slovak Dairy Association – as the representative of milk processors, reached an

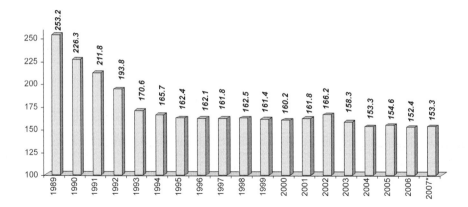

Figure 10. Consumption of milk and dairy products in Slovakia (1989-2007) (Slovak Association of Dairy Farmers).

Table 6. Consumption of milk and dairy products in Slovakia (kg) (Research Institute of Agricultural and Food Economy, 2008).

Category	2002	2003	2004	2005	2006	Estimate 2007
Liquid milk	67.1	63.9	59.1	55.7	55.9	54.5
Cheese and cottage cheese	9.0	9.3	8.2	9.1	9.5	9.6
Milk powder	1.6	1.4	1.4	1.7	1.1	1.5
Butter	3.0	2.8	2.2	2.0	2.0	2.0
Cream	2.2	2.4	2.2	2.8	3.0	2.7
Yogurt	13.1	12.4	12.6	13.1	12.3	12.6
Other dairy products	0.4	0.4	0.3	0.6	1.2	1.7
Total	166.2	158.3	153.3	154.6	152.4	153.3

agreement two years ago. There are different activities, but the most important ones are those aimed at increasing milk consumption.

During the last 3 years several small-scale, but successful activities promoting milk and dairy products were developed and implemented. These events were low-cost and were financed from the budgets of both associations with the active help of dairies supporting these events through their products. After mutual agreement, the associations decided to prepare a system project called 'Promotion and Information Programme on Milk and Dairy Products Consumption in Slovakia'. This project proposal was approved within the Council Regulation (EC) Nr. 2826/2000 on Information and Promotion Actions for Agricultural Products on the Internal Market (EC, 2000). It will be the first project implemented within this EU programme in Slovakia. The duration of the project will be 3 years and the budget will be approximately 4 million Euros.

The project will be implemented by common funds of the EU, national sources and the rest will be financed by Slovak Association of Dairy Farmers (SZPM) and Slovak Dairy Association (SMZ) though a 'Milk fund'. The objective of the 'Milk fund' is to collect financial contributions from the dairy farmers and milk processors. Basically, each dairy farmer and each processor is supposed to contribute one heller (penny) per kg milk to the 'Milk fund'.

Milk balance

In 2007, the total milk balance amounted to 1,413,964.7 t. This was comprised of 973,529 t of raw cows milk purchased from dairy farmers, 412,757.3 t of imported milk, and an additional 27,678.4 t of the stocks (Table 7).

Of this total, 831,046.2 t (ca 59%) were consumed, 550,205 t (ca 39%) were exported and 32,713.5 t remained in stock. There is a positive trade balance of milk and dairy products amounting to 2,260 m SKK. According to the statistics, the import/consumption ratio is just less than half (49.7%) while the export/sale ratio is 57%.

Table 7. Balance of traded cow milk in Slovakia (Research Institute of Agricultural and Food Economy, 2008).

Indicator	Unit	2004	2005	2006	2007
Average number of milking cows	×1000	206.0	198.5	192.5	180.6
Average milk yield	kg per milking cow per year	5,235.7	5,541.8	5,670.1	5,951.4
Production	t	1,078,625.3	1,099,827.0	1,091,737.2	1,074,655.3
Stocks at the beginning of year	t	20,000.0	18,880.2	17,596.4	27,678.4
Milk purchase from dairy farmers	t	950,548.0	974,493.0	970,115.0	973,529.0
Import	t	140,166.0	301,282.0	351,188.0	412,757.3
Total sources	t	1,110,714.0	1,294,655.2	1,338,899.4	1,413,964.7
Export	t	340,334.0	482,192.0	515,042.0	550,205.0
Domestic consumption without natural consumption	t	751,499.8	794,866.8	796,179.0	831,046.2
Stocks at the end of year	t	18,880.2	17,596.4	27,678.4	32,713.5
Import/consumption ratio	%	18.7	37.9	44.1	49.7
Import/sale ratio	%	14.7	30.9	36.2	42.4
Export/sale ratio	%	35.8	49.5	53.1	56.5

Economic efficiency

Economic efficiency of milk production

The economic efficiency of milk production in Slovakia and European Dairy farmers (EDF) countries in 2007 were analysed:
- Economic efficiency of milk production in Slovakia
- Average costs for milk production: 11.02 SKK/l
- Average milk price: 10.07 SKK/l
- Economic result: loss -0.95 SKK/l
Economic efficiency of milk production in EDF countries
- Number of companies analysed: 270 (from 17 countries)
- Average entrepreneurs result: loss €2/100 kg

Relationships among milk producers dairies and retailers

The Slovak Association of Dairy Farmers keeps contact with the dairies and negotiate on the market situation. The common platform is the negotiation between SZPM representing dairy farmers and SMZ representing dairies. Different analyses are conducted to compare prices for producers, processors and retailers and the respective margins. For example, Figure 11 shows the development of producer prices per 1 kg of raw cow milk in comparison with the prices of processor and retailers and their margin for 1 kg of 1.5% skimmed UHT milk.

The volatile pattern, especially in the first half of 2008 is clearly depicted. At the end of 2007, the retail chains implemented a large price increase for dairy products. This caused a drop of 18% in consumption until the end of March. Stocks of milk and dairy products increased both at the level of processors and retailers. Retailers pushed down the processors prices and the processors pushed down the producer prices, but in the end, the consumers prices on the shop shelves remained more or less the same.

The general feature of relations within the dairy chain is the unequal position of the different stakeholders. Retailers abuse their dominant position by pushing down processor prices. Additionally they force the processors to pay listing fees, promotion fees and other hidden payments which are not directly included into the product price.

A law on equal conditions on the market is being framed. This Ethic Codex is under the legislative procedure and should be adopted at the beginning of 2009, but it is unlikely to solve the present situation. It is likely that it will be necessary to adopt a common European law defining the rules for equal position in the market chain including some special rules forbidding market distorting practices by retailers.

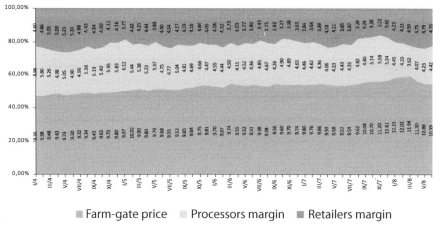

■ Farm-gate price   ■ Processors margin   ■ Retailers margin

Figure 11. Producer price of 1 kg of raw cow milk and processor and retail margin of 1 kg of 1.5% skimmed UHT milk (2004-2008) (Slovak Association of Dairy Farmers).

CAP 'health check'

Slovak dairy sector as a part of the common European dairy sector

Slovakia is a part of the European and world milk market and therefore must analyse the situation objectively. Most of the current problems in the dairy sector are common to all member states. One of the most recent issues of concern is the 'health check of CAP'. The milk quota system was introduced 24 years ago, within a completely different social, economic and political environment. Milk quotas, on one hand ensured the stability of European milk producers incomes, but on the other hand destroyed the 'natural' operation of the milk market.

It is acknowledged that it will be difficult to return the 'artificial milk market system' back to a 'natural milk market system'. It will also be difficult to take the appropriate and sensitive measures which would lead to preservation of a sustainable and competitive dairy sector in the EU as a whole and in the individual EU member states.

It is unlikely that in the current situation, characterised by volatile milk prices, an increase in milk quota is the only possible solution on how to reach a 'soft landing' on milk quotas. It is obvious, that an increase in quota will bring advantages for the 'big milk players' that have the capacity for expansion, but the adoption of this measure may endanger 'small milk countries', such as Slovakia. The Slovak Association of Dairy Farmers will support measures, which will be based on the same business conditions and which will not give preference to either individual production systems or individual member states. Milk producers in the mountainous regions and disadvantaged areas should be given separate sensitive measures.

With regard to the market support measures, the European Commission, on the one hand, presents the growing demand for milk and dairy products on the world market as a big challenge for the European dairy sector, but and on the other hand, it cancelled the export subsidies. Returning to the situation of last year, the increase in prices at all levels of the dairy chain caused the big drop of consumption. At the same time, farm gate prices increased. Suddenly there was over production which affected the internal market and constantly increased the pressure on the whole dairy sector. Finally, there is a question: 'Who will supply the increasing world market demand?' The US Farm Bill, with payments for more cows and a 'feed cost adjuster' partly supplies the answer. World Trade Organisation (WTO) deals definitely endanger agriculture generally and particularly the dairy sector. It is difficult to forecast milk price developments. Constantly increasing costs (feedstuffs, energy, fertilisers, etc.) make the milk market more volatile, while cross compliance and the bioenergy boom do not improve the situation either. The pertinent question is 'Can dairy farmers withstand the evolving conditions and survive?'

References

European Commission (EC), 2000. Council Regulation (EC) No 2826/2000 of 19 December 2000 on information and promotion actions for agricultural products on the internal market. Official Journal of the European Union L 328: 2-6.

Research Institute of Agricultural and Food Economy, 2008. Website available at: http://www.vuepp.sk/

Slovak Association of Dairy Farmers, 2008. Website available at: www.szpm.sk

Cattle sector and dairy chain developments in Georgia, Azerbaijan and Armenia

T. Kartvelishvili

Georgian National Association for Animal Production, 0172 Tbilisi, Georgia;
tamara_kartvelishvili@yahoo.com

Abstract

This study describes the present situation in the Georgian dairy sector and identifies the problems in terms of livestock breeding and forage supply, milk production, its primary handling and subsequent processing. Based on information provided by the Ministry of Agriculture, Department of Statistics and other organisations, relevant data is given, including: number of livestock (milking cows), milk production (industry and households), dairy processing volumes (industry and households), dairy product consumption (on a nationwide scale and annually, per capita), export-import indicators and data on ongoing and completed international projects in the dairy sector. Though it should be mentioned that figures published by state organisations do not adequately show the existing situation of the sectors, consequently making it hard to make sufficient conclusions out of such information. Current milk and dairy production and consumption levels in Georgia, as well as the domestic self-sufficiency ratio are also analysed. For example, the recommended norm for annual consumption is 330 kg of milk per capita, while the average consumption in Georgia was estimated in 2005 at 238 kg and the amount produced by the Georgian dairy industry was just 184 kg. The deficit between consumption and domestic production is presently being filled by imported dairy products. Together, this data illustrates both the current underdevelopment of the dairy industry in Georgia as well as the potential for strengthening and expansion. In analysing the information contained in this document, it becomes clear that the dairy sector represents a critical value-chain in Georgian agriculture and, given that virtually 100% of all milk in the country is produced by farm families, one that is particularly important for rural family incomes as well as for rural development. This article also includes some information about livestock developments in Armenia and Azerbaijan. Azerbaijan is an ancient livestock country. Thorough changes have occurred in the livestock of Azerbaijan during last 12 years. For the purpose of cattle breeds' development, the districts in Azerbaijan are divided into 3 zones: a dairy-production zone, a dairy and meat-production zone and a meat-production zone. The Armenian Agriculture consists of two main sub-branches: agriculture (plant growing) and livestock breeding, which in their turn are divided into various smaller sub-groups. Most agricultural production in Armenia is directed towards crops, which in 2007 accounted for 64% of gross agricultural output. The majority of cattle breeding in the republic is carried out by and based on extensive methods.

Keywords: cattle sector, dairy production, dairy processing

Developments in Georgia

Introduction

Georgia has a vast untapped agricultural potential, in fact, so much that it could increase fivefold its value of crop production. Yet, in order to realise this potential, the nation also faces vast challenges that can slow if not actually prevent the attainment of any goals set for the sector.

Government of Georgia has three broad goals for the nation – economic growth, civil order and poverty alleviation. Now our vision is to become a nation where all Georgians have access to a safe, affordable, nutritious food supply where those who provide food and other agricultural products can do so profitably, safely and with dignity and respect, where the beauty and function of the natural environment is maintained and enhanced and where national security, employment, social and objectives for food and agriculture are met (The Georgian National Food and Agriculture Strategy, 2006).

The ability to develop Georgia's agriculture successfully will be highly dependent on the sector's external factors. The more important of these include economic and demographic considerations, social-political and international factors, and consumer and retail trends.

Economic considerations

Economically Georgia has made considerable progress over the past decade. One of the most significant accomplishments is that:
- real GDP has nearly doubled;
- the real value of exports is up approximately 150%;
- the government expenditures increased nearly fivefold in only ten years;
- inflation is now under control;
- real average monthly salaries have increased;
- all sectors of the economy seem to be growing or have at least stopped any further declines from Soviet era levels(except mining);
- expenditures on education, healthcare and infrastructure have increased exponentially.

There is a reasonable optimism within government that the economic progress of the recent past will continue into the future. In fact, the government feels that real economic growth in the 5.0-7.5% range annually is a realistic expectation for the coming years.

Nevertheless, government forecasts on the economic growth, budgetary approaches and current structure of economy has problem factors. If these factors will influence efficiency of the Georgian economy and the government policy that current optimistic economic prospect it will not be probable to be carried out.

The Russian embargo since 2005 on food and agricultural products from Georgia has significant negative implications for the country. Unfortunately for Georgia, because of the relative sizes of the two economies and given the globalisation of the world economy, Russia can take political actions which affect the food and agriculture sector of Georgia significantly, but do have a minimal or no affect on either the larger country's economy or on its consumers.

Demographic considerations

Demographically three factors stand out in Georgia:
- the national population has continued to decline through 2005;
- the population on average has continued to age;
- natural population increase is approaching zero.

There is a very interesting demographic-economic phenomenon in Georgia that is fairly typical of countries with strong family ties, under high unemployment and no social welfare safety net. In many respects, this interdependency is positive since it helps provide a social welfare safety net as well as providing at least limited capital to the under financed farm sector.

Social-political factors

The government is supporting rural and remote area development in its public pronouncements and discussions with donors. However, these priorities do not seem to be widely held or actively pursued. Some governmental limited reaction in areas where unemployment and high poverty exists takes place, when a natural disaster occurs. These situations cannot totally be ignored. However, in none of these situations has a coherent longer term strategy been build up for identifying and assisting areas with chronic (but not crisis level) poverty or malnutrition.

To date, the primary approach of the government to address existing or potential social or political problems has been to focus on the privatisation of public assets, the development of the energy and roads sectors, the elimination of onerous laws and regulations and the increased funding of education. The apparent hope is that this policy will be sufficiently stimulative such that the economic growth target will be met. This growth in turn would hopefully generate increased employment and improved incomes sufficiently to dampen or eliminate any potential social or political unrest.

In general this approach of Georgian government is not an unreasonable one. The nation has limited resources and virtually unlimited problems or needs, problems and needs which cannot all be addressed equally and simultaneously.

With respect to agriculture specifically and rural development in general, the government has been taking a risk. At this time, agricultural only receives about 1.5% of state budget. Of this, roughly half was provided indirectly by the EU Food Security Program (FSP) budgetary support. Thus, less than half of 1% of the state budget, provided from Georgian revenues, goes to agriculture. This implies a sector that comprises 16-20% of GDP and provides over 50% of employment directly. When production inputs, processing, wholesaling, transportation, government services and other dependent or partially dependent economic activities are also taken into account, plus when other appropriate economic multipliers are applied, it is not unlikely that at least one-third of the total economy and over 55.3% of national employment is dependent on agriculture.

When the economy collapsed after independence, the land was distributed and leased in generally sufficient quantities to enough people to insure that there was not a large portion of the population with any means of support. The structure of agricultural land ownership underwent a significant transformation in the first stage of the agrarian reforms. After the land reforms, about one million households became the owners of nearly 30% of total agricultural land. As a result, over half the working population could support itself at subsistence level through farming. Without this policy, there would likely have been a much larger unemployed urban work force with significant potential for social and political unrest and even violence.

It is unclear whether the continuation of government's approach to the agricultural sector will remain socially and politically viable.

International factors

Georgia is presently part of the WTO. These relationships impose various restrictions that may lock the country into unfair agricultural competition with existing member nations, especially those more developed members with highly subsidised farm sectors. Georgia does not have the resources to provide similar subsidies, but it will not be permitted to levy tariffs to mitigate against such subsidies. On the other had, membership in such organisations does proved certain protections for Georgia when it is trying to control its borders from dumping of substandard or unfairly priced products. It also provides a means by which its exported products can more fairly compete and protected from infringements abroad.

Yet, regardless of whether Georgia is a member of WTO, the EU or any other similar treaty or international affiliations, if it hopes to capitalise on export markets, its agricultural sector must increasingly be able to meet international standards.

Logically Georgia must recognise and defend such laws and standards on products, which it exports. Realistically, this may not be sufficient.

Certain segments of Georgia's agriculture are already highly dependent on export markets, like mandarins, apples, greens, nuts, wine and mineral water. In future, it is expected that this will become increasingly so for other products as well. With exports, one faces not just market uncertainties - but political uncertainties as well. (e.g. Russia)

Key factors of Georgian agriculture

Key factors for Georgian agriculture are production, nutrition and consumption, capital, trade, farms, employment, income, agriculture and market knowledge.

Production

- Nearly 17% of national economy is directly dependent on the agricultural sector and industry. Due to the multiplier effect, possibly more than 30%.
- After declining significantly for nearly a decade, the real value of the food and agricultural sector has begun to increase slightly over the past seven years.
- Livestock production has been slowly but steadily increasing for most categories.
- Crop yield on the average are only 1/3 of their potential.
- Approximately 1/3 of arable land is not in production.

Nutrition and consumption

- Over 60% of consumer income is spent on food (vs. 15-20% in the West).
- Nearly 50% of population consumes less than the FAO minimum recommended level of 2,100 calories per day.
- Over 25% of population consumes less than 1,600 calories per day, which is considerably below FAO's absolute minimum of 1,800 calories.

Capital

- Current capital utilisation in the food and agricultural sector is estimated to exceed 600 million GEL (1 GEL=0.53 EURO).
- Eventual capital requirements for this sector to reach its full potential are expected to exceed 2 billion GEL, it means that shortfall of nearly 1.5 billion GEL.
- During the next ten year, approximately 900 million GEL will be required for the nation to attain its goal for the food and agriculture sector.

Trade

- The reported real value of agricultural exports has been increasingly of importance to the country.
- Food and agricultural imports are 50% greater than exports.
- The country is overly dependent on exports to Russia for virtually all food and agriculture products.

Farms, employment and income

- Georgia has 657,542 farms with an average size of 1.48 ha prior to the next phase of privatisation, after which the average farm size will be 1.70 ha;
- Presently there are 16,000 farms of 4 ha or greater, but these represent 40% of all cropland in private hands (owned or leased).
- About 55% of the national labor force is presently employed in agriculture vs. only 25% in 1990;

Nowadays in Georgia the most agricultural production is produced by household farming, which is oriented on self-provision and is characterised by the low level of production (Table 1). Householders are small and medium output scale and fragmentary.

Small household farms predominate throughout the livestock sphere (Table 2). On average, farmers have 1-9 cows, as well as pigs and poultry. Today farmers possess 1.48 hectares, where they produce vegetables, fruit and grains (maize, sunflower, barley in general) both for their own consumption and for sale. These lands are privatised and are only arable. In this regard the use of such land for pasture isn't appropriate and production of forage grains is small. Livestock tends to graze on pastures which are community property of the whole village. In the evening, the animals go back home and the farmer has to provide additional feed by from stored grain. Very few farmers are able to feed lactating animals properly.

During winter, especially in times of snow, animals are housed in special barns (usually for between 1-3 months each year). During this period the animals are fed by hay and sometimes the diet is enriched with wheat bran, sunflower (in Kakheti Region) or soy-bean (in Samegrelo Region). General animal health suffers greatly due to poor diet during the winter. Advanced dairy farmers possess more animals. These advanced farmers do realise the necessity of improved feeding, rich forage and concentrates for the animals during the winter. Regardless the abandoned pastures and the shortage of the food stocks, development and cultivation of the natural pastures are not yet practiced. 70.7% of the valleys and 95.3% of the pastures are not privatised and are yet under state property - therefore, they are not properly cared for and developed. Production of forage and silage for winter is as well limited due to the lack of appropriate inputs.

Table 1. Total number of holdings and its structure by holding type in Georgia in 2007 (State Department of Statistics, 2007b).

	All holdings	Family holdings	Agricultural enterprises	Other type holding
Number of holding	657,542	656,247	720	375
Structure%	100	99.8	0.1	0.1

Table 2. Number of holdings by size of farm expressed in number of cattle (w/o buffalo) in Georgia in 2007 (State Department of Statistics, 2007b).

1	2	3-4	5-6	7-9	10-14	15-19	20-29
109,513	129,826	106,145	31,938	12,814	6,590	1,399	870
30-49	50-69	70-99	100-199	200-299	300-499	500-999	>1000
519	191	91	38	7	1	1	-

Agriculture and market knowledge

Historically, Georgia is located in a region in the world, where the most agriculturally progressive countries were several generations ago. In present time Georgia does not have effective national research, education, extension or market information systems.

Agricultural Research, Education and Extension (AgREE) systems are practically disorganised and their existence is not meaningful. Unfortunately, Georgian's AgREE system has progressed only minimally towards meeting the needs of the new farming sector. Thus, it must be expected that tremendous challenges will be faced in getting AgREE institutional development started on the right course and once started, in keeping it on track.

The cattle sector

Cattle breeding is the ancient and traditional field of agriculture. In Georgia the indicators of average productivity of cattle were low even during the previous period of crisis in 1989. The average milk yield of cows in all category farms of the country was 1,275 kg, the population of cows was 588,000 and a total of 714,000 tons of milk was produced. During the Soviet period cattle-breeding gave the republic more than half of total livestock production (in currency).

From 1990 on, the transition from the centrally planned economy to the market economy caused crisis that, on its part, resulted in reduction of the number of farm animals, loss of their productivity and loss of the animal production industry (Table 3).

After the vivid decrease in number of animals since 1991, there has been a little, but stable increase process, which has mainly increased number of cows. In Georgia this is conditioned by a relative development of the dairy sector, which is stimulated by government support to farmers for imported cows from EU in 2007.

Most important animal products

The following species are utilised in primary livestock production: cattle, pigs, sheep, goats, poultry, fish and bees. Cow milk is predominant in milk production, while pork and poultry are equally represented in meat production. Locally adapted breeds fulfill a much larger role in all livestock sectors than modern imported breeds. The reason is that high production breeds are simply not imported. Nevertheless, the productivity of the local breeds is low, because the breeds are degenerated. The most important primary livestock products are meat, milk (Table 4), eggs, fish and poultry.

Georgian regions differ in respect of significance of these products. The importance of these products relates to particular regions, depending on geographic, social and economic status and management of natural resources.

In last years there has been a significant reduction in livestock products export, since we are not self-sufficient in livestock production (Table 5). Social difficulties and privatisation processes have substantially contributed to the decline in livestock production.

Table 3. Changes in of cattle population in Georgia in the period 1996 till 2007 (×1000) (State Department of Statistics, 2007a).

Cattle			Of which cows		
1996	2004	2007	1996	2004	2007
973.6	1,242.5	1,128.9	551.7	728.0[a]	571.4

[a] The number of cows has sharply fallen in 2005-2006 and then increased again.

Table 4. Livestock production in Georgia in 1996 till 2007 (×1000 tons) (State Department of Statistics, 2007a).

	1996	2000	2004	2007
Meat	117.8	107.9	109.2	111.2
Milk	530.3	618.9	780.4	857.6
Eggs[1]	350.2	361.4	496.6	383.2
Wool	3.0	1.9	2.2	2.2

[1] Number of eggs ×1,000,000.

Animal health

Generally human health is directly affected by what the nation does or does not do with respect to livestock diseases. Some of these can be spread with severe health effects for humans, e.g. anthrax, tuberculosis. Most recently, a new problem has surfaced that of avian influenza and swine plague. Because of the potential problems for human from these livestock diseases, government cannot simply take a passive role - cannot rely solely on individual farmers to take the proper steps to monitor, treat for or eradicate such problems, farmers who generally have little money for medicines or vaccines or easy access to them. In addition, even when producers, such as commercial poultry farmers can actually treat certain livestock health problems, another human health problem can arise. This area cannot be left to the private sector to do all that is necessary to insure such problems do not surface. Presently government has a system in place for vaccinations only in cattle, but there is no tracking and recording system whereby it can be determined when the last vaccination was made. (Unfortunately the Government of Georgia cancelled all state programs related to epizootic diseases like FMD, anthrax, rabies, tuberculosis, brucellosis, etc.).

Fortunately, while there are shortcomings in the current system for controlling livestock diseases that might affect human health, there have been not problems of significance to date. This cannot be relied on to continue indefinitely (avian influenza and swine plague is a perfect example).

Most problems can be addressed by the affected farmer taking appropriate control measures. However, there can be problems in livestock where on-farm control is not adequate even with preventive measures being taken.

Food safety

In 2005 Georgia still had a Soviet style food control system which did not work to protect consumers or industry. Rather it was applied mainly as a mean of supporting a large network of inefficient and ineffective inspectors and laboratories. The food safety system focused on end product certification and control. Within the system at that time, inspectors' technical knowledge and skills were outdated (same situation till today).

Table 5. Share of livestock in agricultural output in Georgia in 1996 till 2007) (State Department of Statistics, 2007a).

	1996	2000	2004	2007
Livestock in % of total agricultural output	0.45	0.52	0.50	0.52

The law on Food Safety and Quality was adopted in December 2005 by Parliament of Georgia (Ministry of Agriculture, 2005). The new law establishes a legal framework consistent with WTO requirements and the EU *acquis communautaire*.

The launch of the reform in the food safety system coincided with a major drive by the new government to deregulate the economy, which heavily influenced the process as to what was perceived as the appropriate extent and nature of official control and regulation. Components in the Food law draft relating to licensing of food establishments regarding internal safety control systems and traceability were considered to provide opportunities for continued lawful extortion of 'fines' from food businesses.

Approval of the law on Food and Quality is just the first step in the reform process. There is no value to the law, no matter how good, unless it is properly implemented and enforced. According to decision of the Parliament of Georgia the law is suspended till 1 January 2010. Thus, there is still a considerable distance to go before Georgia will have an acceptable level of food safety.

Farm animal genetic resources

Nowadays the genetic resources of local cattle in Georgia are represented by three breeds: Georgian Mountain Cattle, Megruli Red and Caucasian Nut Brown, and also Georgian buffalo (Saghirashvili and Karttvelishvili, 2006; Saghirashvili *et al.*, 2006)

Georgian Mountain Cattle

This breed is one of the oldest breeds, first of all for the production of milk. It is also used as beef cattle and draught force. During Soviet period the number of Georgian Mountain Cattle in the social sector composed 16.2% of the cattle total population. Presently it is preserved on the Southern slopes of Caucasus mountain range. The extension zones of this breed are rich of rivers and brooks heads, we rarely meet plains. In the most zones of extension the bent of pastures reaches 30-35° and other cattle could not use it, except Georgian Mountain Cattle.

Georgian Mountain Cattle is very small, the height in wither of the cow is on average 98-100 cm. It is characterised by low milk yield in the conditions of primitive feeding, but in the case of improved feeding and care-keeping the milk yield increases on average to 2,000 kg with 4.2% butterfat (fatness). During the increase of milk yield, Georgian Mountain Cattle maintains fat percentage composition in milk. It has a hard constitution, endurance, milk butter-fat and high culinary peculiarities of meat. The most part of milk production is realised during the pasturage period, but after taking the cow on stationary feeding, milk yield reduces quickly and stops.

Megruli Red Cattle

Megruli Red Cattle represents the breed of universal usage. It is raised with the completion of local small-body cattle by farmers in 60s of 19th century. Megruli Red cattle spent summer in alpine zones of mountains, but in winter it is pastured in Kolkheti bogs without stationary and supplementary food. In nomadic conditions the milk yield of these cows was increasing from 2-3 to 7-10 liter. This breed is permanently in the open air, so this factor conditioned its adaptability towards local conditions, health endurance, hard constitution and good working peculiarities. The constitution of this breed is mostly towards the milk production herd.

Caucasian Nut brown (grey) breed

This breed is one of the most significant achievements of the zoo-technical science in the 20th century: it meant the establishment of the Caucasian Nut Brown breed on the basis of joint work of the Caucasian countries' scientists. This breed is raised by crossing of Georgian, Armenia, Azerbaijan and Dagestan local cattle mainly to the Swiss Brown breed.

Unfortunately, consequently to the minimum level of feed and care-keeping in intensive from was achieved in Georgia, it was practically impossible to raise more productive breed. The Caucasian Nut Brown breed composed more than 90% of the total cattle livestock population in Georgia during the existence of social farms: the breed was economically justified. For example: there was 1.1 million livestock according to 1990 data, of which 330,000 cows. The annual milk yield was 2,400-2,800 kg in the conditions of average feeding with 3.8-4% fatness. With improved conditions of feed and care-keeping milk yield reached 3,500-4,500 kg, while the maximum milk yield was 8,789 kg, which indicates high genetic abilities of this breed.

Nowadays the amount of Caucasian Nut Brown breed exceeds 95% of from the total cattle population in Georgia, but their productive indicators do not correspond to breed standards. This is caused by the fact that breeding farms do not function, zootechnical registration is out of order and artificial insemination does not exist. There are zero breeding farms in the country, which finally will cause the degradation of the Caucasian Nut Brown breed. According to our data, an analogous situation concerning the Caucasian Nut Brown breed exists in Armenia and Azerbaijan.

Georgian buffalo

Buffalo breeding has a long history in Georgia. In South Caucasian Countries, in 1960 buffalo population was more than 500,000, but then it declined and in 2007 was fixed at 29,541 in Georgia. The main part of milk production is received during pasturage period, but after taking the buffalo-cow on stationary feeding, milk yield reduces quickly and stops.

Buffalo breeding is directed towards work-dairy-meat. For the dairy part it's nearly the same as for the local cattle breeds. The buffalo gives 1,300-1,500 kg milk with nearly 7, 8% fatness. Buffalo dairy productivity has the potential of 3,000 kg.

Georgian buffalo is like the Armenia and Azerbaijan buffalos in constitution, which is caused by the closeness of their extension area, common origin and similarity of care-keeping conditions.

Dairy production and processing

Milk production

Cattle husbandry is mainly concentrated in private farms throughout Georgia (Table 6). Accordingly, 99.9% of the milk production falls in this sector. Enhancement of quantity, quality and energetic properties of animal feed will stimulate the production of all livestock products, including milk.

Georgia has a more than suitable natural and climate conditions for cattle husbandry. Today the milk production volume is steadily increasing, although the data varies by region and generally low productivity per cow is still a major problem (Figure 1).

Improvements to increase the productivity of livestock breeds, meadows and pastures, stimulate cultivation of the food crops, introduction of effective technology related to animal indoor feeding, will further increase the stable production of high-quality milk during summer and winter.

Table 6. Milk productivity and milk production by farms of all categories in Georgia in 2004 till 2007 (State Department of Statistics, 2007a).

	2004	2005	2007
Total milk production (×1000 t)	780.4	787.7	857.6
By households	779.6	786.7	856.8
By agriculture enterprises	0.8	1.0	0.8

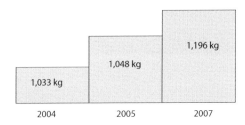

Figure 1. Milk production level in Georgia from 2004 till 2007 (average annual yield per cow in kg) (State Department of Statistics, 2007a).

Dairy processing industry (Millennium Challenge Georgia Fund, 2006)

Owing to the positive economic-political status, appropriate investment policy and simplification of the tax system, the local production of milk products has been increased during the last few years. In 2005 the processing volume of milk and milk products was estimated at 98,616 tons. Dairy processing in Georgia is done on three levels:
- farmers/households processing milk from their own herds;
- small scale cheese producers processing fresh milk collected from adjacent farms;
- large scale industrial dairies, processing predominantly imported milk powder as well as locally collected milk.

It is common for Georgian rural householders to process milk at home and sells various dairy products in the regional or central markets. In 2005 rural households produced 89,251 tons of dairy products. Most of fresh milk is transformed into cheese or matsoni (yoghurt type) by the farmers in their houses (Table 7). Transforming fresh milk into cheese and matsoni extends the marketing window for the dairy products and allows farmers to trade with the value-added goods. Most cheese is sold as unbranded large block. The retailers cut them at the time of sale. There is some linkage between cheese makers at the village level and traders, buying large quantities at the farm gate or at the market place. However, in general the dairy sector is characterised by its lack of formal structure and of the obvious channels of farmer to dairy and dairy to retail outlet.

Table 7. Dairy products processed by households in Georgia in 2005 (State Department of Statistics, 2005b).

	Matsoni	Sour cream	Cottage cheese	Cheese	Curd	Butter
Total (tons)	34,587	256	1,508	49,095	4,372	2,433

There are a number of small-scale cheese producers throughout the country with the capacity of processing one to five tons of milk per day. They collect milk from their neighbours; produce Sulguni cheese (Mozzarella type) in a basement of a village house adjusted to such processing operation and sell cheese in Tbilisi or in a central town of the region.

These small processors usually employ 5-8 people; however they are operating without proper registration and certification. Their production is subject to seasonal fluctuation and stops or sharply drops in the winter season when most of cows dry off. Shortage of milk stimulates an increase of its price and makes collection more expensive.

At the same time, these factors drive the cheese price up, so that those processors who stay in operation during winter season can maintain viability. The small dairy units usually have a few pigs and feed whey to them. This is an advantage over the centralised larger dairies which waste whey and are required to conduct its additional treatment before discharging it into a sewage system.

It is noteworthy that in 2003 the data related to processed milk and dairy products did not exceed 5,559 tons. In 2005 the production of milk and dairy products had gone up to 9,365 tons (Table 8). Regardless the numbers of problems existing within the agricultural sphere of Georgia; production of industrial food products (milk and dairy) has increased by 1.7 times.

Today, a number of small and medium-scale enterprises and several large-scale milk processing plants are functioning in Georgia. Large and medium-scale dairy plants are located in Tbilisi. They have made considerable investments into their processing and packing equipment and keep the product quality high. The lack of the appropriate cooling tanks and refrigerated trucks, poor condition of rural roads and the fragmentation of dairy farms inhibit the collection of raw milk from regions to the processing plants. Products, principally matsoni, milk, sour cream and cottage cheese, are sold to supermarkets and small shops mainly in Tbilisi and other cities. Dairy factories' production is mainly based on reconstituted powdered milk. Only few of them are processing raw milk, although significant efforts are made by some processors to incorporate local natural milk in their products. This tendency is further supported by economic considerations. As processors specify, production on dry milk costs more than purchase and transportation of fresh milk the remote regions in the summer which are on two-three hours drive distance from Tbilisi.

Milk collection drops in winter time when most of cows dry off and those who don't - drop milk yields by half due to poor feed and temperature stress. Reduced supply of raw milk drives therefore most of large dairy plants cease raw milk collection in winter. Pursuant to data from the Department of Statistics (in year 2005), the largest share in milk products represented imported milk and milk products: cream, concentrated milk and milk powder.

Table 8. Dynamics of dairy production industry in Georgia in 2003 till 2005 (tons) (State Department of Statistics, 2005a).

	2003	2004	2005
Fresh and processed liquid milk and cream	721	940	1,067
Cheese and curds	323	412	559
Butter	284	682	1,159
Matsoni and other products	1,643	2,190	3,075
Other milk products	2,588	3,256	3,505
Total	5,559	7,480	9,365

Milk and dairy products consumption

Georgia has only now obtained the possibility to produce more product than it is consumed locally. Neither the appropriate input nor the value chains (starting from the agriculture, passing through the processing and trade cycles and addressing to the final consumer) are available yet. In addition to that, the quality and safety level of the food product does not correspond to the standards that are desired by the processing industry and the final consumer. These problems are still painful due to shortage of funds, unavailability of the production input, etc. These obstacles hinder the development of the agricultural sector.

In comparison with the previous years, in 2005 the consumption of milk and other dairy products decreased due to an increase in price ranging from 50%-150% (Table 9).

It should be also noted that per capita consumption of milk and dairy products has been decreased from 242 to 238 kg during 2003-2005 (Figure 2). During the same period, the local per capita production of the dairy products amounted to 176 to 184 kg. The self-sufficiency ration varies between 71% and 75% accordingly. According to the physiological norms, the rational per capita consumption of milk and dairy products should total to 330 kilogram annually in Georgia. Therefore, it will be the best to stimulate the production and consumption of these products.

Table 9. Milk and dairy products consumption by household of Georgia (in tons) (State Department of Statistics, undated).

	2003	2004	2005
Fresh and processed liquid milk and cream, milk powder	230,800	228,600	228,600
Cheese and curds	46,300	45,100	42,900
Butter	7,000	7,200	7,200
Matsoni and other products	37,200	37,700	39,300
Other dairy products	5,300	5,400	5,500
In total	326,600	324,000	323,500

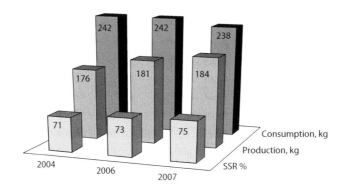

Figure 2. Dairy production (kg), consumption (kg) and self-sufficiency ratio (SSR) (%) (State Department of Statistics, undated).

Import of products

To satisfy the high and steady demand on range of high-quality and diverse dairy products, most of these products are imported to Georgia. To improve the cattle breeds, 8,000 heads of cattle are delivered into the country in 2004-2008 (Custom's Department data). Import of milk and dairy products since 2003 is increasing (Figure 3).

Export of products

In line with the data retrieved from the State Department of Statistics as of 2008, the share of cattle export is very insignificant within the overall export. In 2006 and 2007 the share equaled to zero. Milk and dairy products export is less as compared with the data of 2004 (Figure 4).

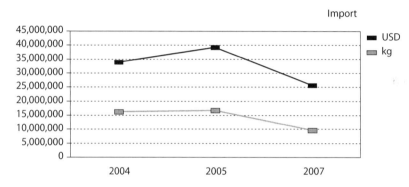

Figure 3. Dairy product import into Georgia in 2004-2006 (kg, US$) (Custom's Department, 2006).

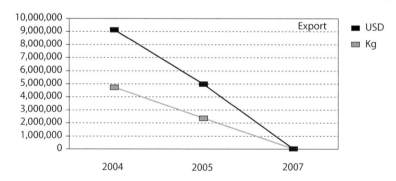

Figure 4. Dairy products export from Georgia in 2004 till 2006 (kg, US$) (Custom's Department, 2006).

International projects

During the last few years several international projects were implemented in Georgia. The projects supported development of milk and milk production sector and explored the stable markets thereof. The most important was the 3-year project implemented by SIDA that was completed in October, 2005. Under the project, the special trainings were conducted to refine the skills of farmers and milk producers, finally leading to stimulation of milk production and collection. During the completion period, the project was mainly oriented on milk quality and production system. Under the project, the informational and marketing services were provided.

During the implementation period of Scanagri project 'From cow to consumer', two milk collection centres were established in East Georgia. The project also carried out a marketing program which included farmers training, capacity building, raising of public awareness about milk consumption through advertising campaign, etc. Despite mentioned results, upon the completion of the project terms, some tasks of the project remained undone which played the decisive role in the selection of OPTO International as the implementing organisation of the second phase of SIDA assistance in Georgian dairy sector.

From 2005, OPTO International is delivering the project 'Dairy sector support in Georgia' into different regions of Georgia. The project supports the small and medium-scale milk production enterprises and dairy market. As of today, six milk collection and two milk production enterprise mainly focused on production of cheese and Matsoni have been founded under the project. The project will provide assistance in terms of overall equipment and output sale; namely, OPTO has an agreement with the large-scale milk producer enterprises on purchase of the milk from the collection centers. Today the equipping process is underway. In addition to that, the project has established a milk collection centre in South Ossetia.

Conclusions for Georgia

Pursuant to the trends described above, for developing the dairy sector it is necessary to focus on the following facts and directions (advices partly derived from '*Dairy production and processing in Georgia, 2006. Millennium Challenge Georgia Fund*'):
* Nearly 17% of national economy is directly dependent on the agricultural sector and industry; due to the multiplier effect, possibly more than 30%.
* After declining significantly for nearly a decade, the real value of the food and agricultural sector has begun to increase slightly over the past seven years.
* Livestock production has been slowly but steadily increasing for most categories.
* Crop yield on the average are only 1/3 of their potential.
* Approximately 1/3 of arable land is not in production.
* Improvements to increase the productivity of livestock breeds, meadows and pastures are needed: stimulating cultivation of food crops, introduction of effective technology related to animal indoor feeding, and increase of a stable production of high-quality milk during summer and winter.
* There is a high and stable demand on milk and dairy products in the country.
* The sector has solid potential of development to replace the imported dairy products. In total 323,500 tons of milk and milk products are consumed in Georgia annually, while production of the industrial output totals to 98,616 tons, therefore, to satisfy the local demand an additional 224,884 tons of milk and dairy products are required.
* The small and medium-scale livestock farms should be consolidated and enlarged and the packaging-storage facilities should function in order to explore and obtain stable markets for milk and dairy value-added products; moreover, new technologies should be introduced and the enterprises modernised.

- A centralised marketing network should be created thus assisting the farmers and entrepreneurs to uninterruptedly supply the milk and dairy value-added products.
- Milk collection from many small farmers is associated with huge expenses, thus making the price uncompetitive for the producers. Therefore, processors prefer to deal with the large-scale farms and/or utilize imported milk powder. Establishment of the milk collection centers will benefit not only the small farmers but the value-added enterprises as well.
- The processing enterprises should be established in the districts specialized in dairy farming.
- In terms of food safety, the documentation confirming the appropriate quality should be attached to the product. In order to produce a sound product, the veterinary service centers should operate in each region. The best scenario is if they will operate within the Farm Service Centers. The milk and dairy products should pass the laboratory analysis.
- The value chain should be formed in such a way that it starts from the basis (farms), passes through the processing industry and trade and is addressed to the final consumer.

Developments in Azerbaijan

Introduction

One of the main goals for Azerbaijan today is to reduce the dependency of the economy on oil and assure an expansion of economic development to the rural areas. Being the third biggest sector in the Azerbaijani economy after oil and construction, agriculture possesses the biggest share of employment (in 2006, 39.1% of total employed population was working in agriculture and only 1% in the oil sector). Agriculture has also a huge influence on poverty reduction in rural areas.

As result of the agrarian reforms implemented since 1995, market relations have been established in the domestic economy, land and property are effectively used, the field structure of the agrarian sector has improved, entrepreneurship has improved, and the appearance of villages has changed. Today the number of agricultural producers is 1,208.700, of which 99.98% is comprised of private, and 0.02% of public, farms. In addition, there are 78,648 cottage farms dealing with production, processing and sale of agricultural products. In total, 66.8% of specialised agricultural producers are family-villagers, 32.8% are housekeepers and 0.2% are farmers.

In 2007, agriculture, hunting and forestry accounted for 7.1% of gross domestic product (GDP). In 2007 Investment in agriculture, hunting and forestry increased by 42% compared with 2005. The budget allocation to agriculture, fishery and forestry increased by 3.5% in comparison with 2005 to 37.2%. The total area of land owned and rented by agricultural producers is 2,324,200 ha (69.3% family-villager, 11.4% housekeeping and 2% farming). Each agricultural producer has on average 1.92 ha of land area.

Notwithstanding the positive results of the last year, there are still several problems in the agrarian sector, such as in the field of cattle breeding (weak artificial insemination stations low-level of supply with equipment, seed, liquid nitrogen and special vehicles). Fundamental actions are needed to improve the feed base, to minimise the delay in implementation of a set of actions to improve private poultry farms and poultry factories, to assure proper veterinary control of animal production, processing, procurement and in import of animal products, and improving the reproductive efficiency of animals.

Cattle sector

Azerbaijan is an ancient livestock country. At present, 99.5% of cattle and 100% of poultry are in farmers' and cooperatives' hands and only 0.5% of cattle is in the ownership of the state companies (Figure 5).

Figure 5. Structure of holdings in Azerbaijan in 2007 (AnGR NC, 2008b).

After the agrarian reforms had taken place, a steadily growing number of cattle and poultry can be observed. The production of all types of livestock increased appreciably. On 1 January 2008, there were 2,512,000 cattle in the Republic, including 1,215,000 cows and buffalos (Table 10). The relative density of breeding herds is 47.8% from general stock, whereas in the Soviet period this parameter did not exceed 26 %.

Now, the number of cattle and small ruminants considerably exceed even the highest numbers, in previous times. Today there are on average 2.3 cattle, including 1.1cows and 7.5 sheep in each family. Last year, production amounted to 294,000 t of meat live weight, 1,341,000 t of milk, and 871 m of eggs (Table 11). According to the Food Safety program, the per capita consumption in Azerbaijan was, on average, 22 kg meat, 179 kg milk and 97 eggs. Of course, this is far below the recommended nutritional level of consumption of livestock products.

The growth of production of livestock basically happens in a natural way. The number of cattle has increased so much that there are great difficulties in maintenance of pastures because of overgrazing. Milk yield per cow is a little over 1,130 kg.

Due to production problems from 1990s on, breeding of cattle declined considerably. During the transitional period of agrarian reforms, the cattle breeding system was let drift and no other work was carried on. For the purpose of cattle breeds' development, the districts in Azerbaijan are divided into 3 zones: a dairy-production zone, a dairy and meat-production zone and a meat-production zone (Genetic Resources Institute of National Academy of Sciences of Azerbaijan, 2006).

Table 10. Changes in of cattle population in Azerbaijan in the period 2004 to 2007 (×1000) (AnGR NC, 2008b).

Cattle			Of which cows		
2004	2006	2007	2004	2006	2007
2,293.6	2,445.2	2,512.2	1,007.5	1,184.1	1,215.7

Table 11. Livestock production in Azerbaijan in 2006 and 2007 (×1000 t) (AnGR NC, 2008b).

	2006	2007
Meat	274.2	294.5
Milk	1,300.6	1,341.2
Eggs[1]	761.6	871.0
Wool	13.6	14.0

[1] Number of eggs ×1,000,000.

Farm animal genetic resources

There are 27 breeds of cattle in Azerbaijan, two of them are local breeds: Caucasian Nut brown (grey) breed and Red Kazakh cattle, besides there are 2 breeds of buffalos (one of them is local) and local breed of zebu - Azerbaijan Zebu (Genetics Resource Institute of National Academy of Sciences of Azerbaijan, 2006).

Red Kazakh Cattle

This is the product of local selection. Its color is golden-red, sometimes dark-red. Kazakh cattle are robust and are of good conformation. The udder has a regular shape, and the animal is resistant to various diseases. It has good stamina, is tolerant to harsh local conditions of management and has relative high fertility. The milk yield of cows per lactation is 1,900-2,000 kg with a fat content of 4.2-4.6%. The average live weight of cows reaches 380-400 kg and bulls weigh 450-500 kg. Calves give over 50% meats.

Azerbaijan Zebu

The Azerbaijan Zebu is one of the rare species of animals found in the Republic. Its colour is black and dark-brown. You can see a hunch on its back. Azerbaijan zebu and its hybrids are very robust and suited to local management conditions and resistant to diseases. They are well adapted to local management conditions which have reinforced and perfected certain biological features including their precocity. The Azerbaijan Zebu reaches 300-350 kg live weight and gives about 58.8-60% of its weight as meat. The Zebu cow gives over 500 kg of milk per lactation with 5-6% of fat.

Azerbaijan Buffalo

As a result of a long and intensive genetic selection and livestock breeding process, together with the creation of good and beneficial conditions for feeding and managing the cattle, the Azerbaijan buffalo's quality potential has been significantly improved. The production of a female buffalo is 1,300-1,500 kg milk per lactation with 8-12% fat. The average live weight of the buffalo reaches 400-500 kg for cows and 800-1000 kg for bulls.

Conclusions for Azerbaijan

- Cattle husbandry is the most significant field of agriculture in Azerbaijan.
- In Azerbaijan the purposeful agricultural reforms have created a real potential for reorganising the pedigree livestock sector in conformity with the new economical conditions.
- The livestock production in Azerbaijan meet people's demand in the internal market.
- In Azerbaijan, the well adapted breeds to local condition and rational breeding by the farmers are strong points for a good development of the cattle sector.
- A weak system of artificial insemination is a big problem in the cattle breeding sector in Azerbaijan.

Developments in Armenia

Introduction

Agriculture is carried out mainly in the valleys and mountainsides of Armenia's uneven terrain, with the highest mountain pastures used for livestock grazing. Only 17 percent of the country's land is suitable for farming, which severely limits agricultural production. Fertile volcanic soil allows cultivation of wheat and barley as well as pasturage for sheep, goats, and horses. Despite the limitations of this sector, agriculture provides the largest source of income for Armenia.

Agriculture in post-Soviet Armenia reflects the results of the privatisation of collective farms, the distribution of that land to the workers, and the large unemployment resulting from closed factories. Gross agricultural production has not changed much since 1990. Arable land and orchards, previously farmed as a unit dedicated to one crop, now produce a variety of products. Land holdings by rural Armenians range from one to three hectares on average, often comprised of smaller, non-contiguous parcels. As a result, farming these small plots is very inefficient. On the bright side, Armenia is blessed with multiple agronomic zones, conducive to production of a range of crops and animal products. There are extensive high meadows suitable for goats and sheep, which can supply the milk for a variety of good cheeses. Agricultural production is heavily biased toward crops, which in 2007 accounted for 64% of gross agricultural output.

The agricultural sector remains an important contributor in Armenia. Not including food processing, which is factored into industrial output, the agricultural sector provides 30-35% of GDP in an average year. Including food processing would raise the total sector contribution to GDP to nearly 45%. The sector is also a major employer with over 40% of the population dependent on the agricultural sector for employment. However, according on the USAID, USDA and Ministry of Agriculture of Armenia (2006) team report there are significant constraints to the development of the agricultural sector in Armenia. These impediments to a thriving and efficient agricultural sector include:

- lack of adequate transportation and the high cost for transportation;
- structural financial market impediments causing a lack of suitable financial instruments: insufficient credit available for long term investments and high interest rates;
- the relative high cost of inputs at both the farm and the processing levels;
- shortage of experienced managers;
- lack of governmental support (including a lack of applied research information);
- small farms resulting in little or no efficiency of scale.

Cattle Sector

In Armenia at present, a high percentage of cattle are in farmers' and cooperatives' hands. From 1990 on, the trends of reduction in number of animals on a farm, loss of their productivity and loss of the manufacturing industry of animals have been stopped. After the agrarian reforms had taken place, a steadily growing number of cattle and poultry can be observed (Table 12), comparable to a similar trend in the other two south Caucasian countries- Georgia and Azerbaijan.

Main animal husbandry products manufactured in the republic are milk, beef, chicken, pork, eggs and honey. Wool and leather are regarded as important raw material. These products have different significance for different social segments. Population in the alpine zones is engaged in livestock (cattle and sheep breeding); in lower zones pig and rabbit breeding and apiculture are added to them. Fowls are raised mainly in areas located in the proximity to urban areas. Production of meat, wool and milk are supported by local conditions and well adjusted breeds, while production of eggs and poultry is based on highly productive breeds of fowls, intended for giving chicken meat and eggs.

Table 12. Changes in of cattle population in Armenia in the period 2004 to 2007 (×1000) (AnGR NC, 2008a).

Cattle			Of which cows		
2004	2006	2007	2004	2006	2007
397.1	478.7	483.3	220.9	262.1	277.5

Livestock production in 2007 reached 66,800 tons of meat (slaughter weight), 620,000 tons of milk, and 464 million eggs. But only milk production increased significantly during the post-Soviet period.

Farm animal genetic resources

The main cattle breed in Armenia is the Caucasian Grey/Nut brown Breed (the same as in Georgia and Azerbaijan). Secondly, there is also the Black-and-White breed (Rukhkyan *et al.*, 2005).

Black-and-White cattle

Import of this breed was dictated by the necessity of moving livestock to stationary conditions, which occurred particularly in regions closely located to urban areas. From 1970 on, many female species of this type were brought into Armenia and many Black and white herds were founded. They had different productivity indices than local cattle. Under favourable conditions of feeding and care their milk productivity was between 4,000 and 5,000 kg, but it went down under unfavourable conditions. Therefore raising this breed is not expedient. Now this breed is considered as second important cow breed in the country. They do not adjust well to mountainous pastures' conditions. Rural farmers wishing to acquire this type of cow should have in mind that they should be kept in stationary conditions or in pastures close to households. Only female heifers can be taken to mountainous pastures given they are not stony. Their average live weight in Armenia is 480 kg at first calving, 520 kg at second and 550 kg at third. Milk yield in 305 days in 1st, 2nd and 3rd lactation is 3,250 kg 3,600 and 4,000 kg, respectively. Fat is 3.6% and protein is 3.2%. Semen of bulls can be obtained from pedigree farms in Russia or stations specialised in artificial insemination.

In order to raise the efficiency of this type of cows they are crossbred with Holstein breed bulls. The new breed is noteworthy for its increased live weight, high milk yield and productivity in conditions of industrial raising. But the use of semen of Holstein bulls cannot be efficient if the developed generation is not kept with care and in favourable conditions.

Conclusions for Armenia

* Agriculture will remain a very important agriculture sector in Armenia. Agriculture will not be the engine for growth over the long term. Armenia's agricultural production capacity is limited.
* Livestock products produced in Armenia can not meet the national demand.
* A main problem in the livestock sector in Armenia is the weak cattle breeding system.

References

The Georgian National Food and Agriculture Strategy, 2006. Food and Agriculture National Strategy 2006-2015, Tbilisi, Group of Experts, Third Project, 2006.

Millennium Challenge Georgia Fund, 2006. Dairy Production and processing in Georgia. Agribusiness Development Activity (ADA). Available at: www.ada.ge/files/103_137_652293_DairyProductionEng.doc

Saghirashvili, G., Kartvelishvili, T, Kishmareishvili, N. and Tsurtsumia E, 2006. Developments of cattle husbandry in Georgia, farm management and extension needs in Central and Eastern European countries under the EU milk quota. In: Kuipers, A., Klopcic, M. and A. Svitojus (eds.), EAAP Technical Series No 8, Wageningen Academic Pubishers, the Netherlands.

Saghirashvili, G. and Kartvelishvili, T., 2006. Georgian native domestic animal breeds.

Custom's Department, 2006. Yearbook. Georgia Customs Department.

State Department of Statistics, undated. Websitie available at: http:///www.statistics.ge/index.php?plang=1

State Department of Statistics, 2007a. Georgia Agriculture Statistical Abstract 2007. Georgia State Department of Statistics.

State Department of Statistics, 2005a. Georgia Agriculture Statistical Abstract 2005. Georgia State Department of Statistics.

Ministry of Agriculture, 2005. The law on Food Safety and Quality. Available at: http://www.maf.ge/?class=3

State Department of Statistics, 2007b. Agriculture Census of Georgia 2007. Georgia State Department of Statistics.

State Department of Statistics, 2005b. Households of Georgia 2005. Georgia State Department of Statistics.

Genetic Resources Institute of National Academy of Sciences of Azerbaijan, 2006. National report on the state of the animal genetic resources in Azerbaijan. Genetic Resources Institute of National Academy of Sciences of Azerbaijan, Baku.

AnGR NC, 2008b. Report of AnGR NC of Azerbaijan. Conference material for: Monitoring for future effective management of farm animal genetic resources in Caucasus region/Armenia, Azerbaijan, Georgia/and Kazakhstan. European Regional Focal Point for Animal Genetic Resources (ERFP) & Georgian National Association for Animal Production (GNAAP) by support Georgian State Agriculture University, Bazaleti, Georgia 10 July, 2008.

AnGR NC, 2008a. Report of AnGR NC of Armenia. Conference material for: Monitoring for future effective management of farm animal genetic resources in Caucasus region/Armenia, Azerbaijan, Georgia/and Kazakhstan. European Regional Focal Point for Animal Genetic Resources (ERFP) & Georgian National Association for Animal Production (GNAAP) by support Georgian State Agriculture University, Bazaleti, Georgia 10 July, 2008.

Rukhkyan, L., Gasarjyan, N. and Chitchyan T., 2005. Country report on the state of the Armenia's animal genetic resources. Ministry of Agriculture, Armenia, Yerevan 06 June 2003.

USAID, USDA and Ministry of Agriculture of Armenia, 2006. Armenia Agriculture Assessment. Team Report.

Cattle sector and dairy chain developments in Poland

J. Fałkowski[1], A. Malak-Rawlikowska[2] and D. Milczarek-Andrzejewska[1]

[1]*University of Warsaw, Faculty of Economic Sciences, Długa 44/50, 00-241 Warsaw, Poland; jfalkowski@wne.uw.edu.pl;* [2]*Warsaw University of Life Sciences, Faculty of Agricultural Economics, Nowoursynowska 166, 02-787 Warsaw, Poland*

Abstract

In the last decade the dairy sector in Poland experienced thorough and dynamic changes. Considerable innovations were observed with respect to production and marketing practices at all stages of the food-supply chain. The institutional environment, in which all economic agents operate, changed as well. All this affected important relationships between dairy producers and processors. Against this background, this paper briefly presents the main developments and adjustments that took place in the local dairy sector. The analysis is based on the results of qualitative and quantitative research conducted in 2006 and 2007, respectively.

Keywords: dairy sector, Poland, vertical coordination

Introduction

The overthrow of the communist system in 1989 and the adjustments to a market economy drastically affected the socio-economic environment in Poland. The agricultural sector was no exception. Particularly spectacular changes took place in the dairy sector. This is illustrated by the fact that, since the beginning of 1990's, the number of dairy farms decreased by more than one million. Other changes include decreases in the number of processing enterprises, necessary quality improvements and increased efficiency of milk production. Important changes occurred also in the institutional environment, in which all economic agents operate. These comprised changes to both international regulations and domestic policies. As an illustration of the former, one can mention, for instance, decisions made under the auspices of World Trade Organisation or regulations adopted within the European Union. As far as domestic policies are concerned decisions concerning forms and amounts of subsidies, or the general approach to the agricultural sector and rural areas, serve as examples (Wilkin *et al.*, 2006).

Taking into account the above-mentioned phenomena, this paper aims to describe in detail the main developments that occurred in the dairy sector in Poland in the last decade, the changes within the production sphere, and results of an analysis of the dairy sector from the dairy supply chain perspective. Special attention is paid to milk marketing and cooperation between dairy processors and milk producers. The information presented draws on several sources of data. First, it uses data collected by the Polish Statistical Office. Second, it takes advantage of qualitative research conducted in 2006 (Wilkin *et al.*, 2006). This research included 36 semi-structured interviews with experts and representatives of all stages of the dairy food chain. In addition, it draws on the information collected during 5 focus group meetings with farmers. Third, the paper uses quantitative data from a survey conducted among 397 farms in 2007 (Milczarek-Andrzejewska *et al.*, 2007). Both these research initiatives were focused on two regions located in north-eastern Poland, namely Warmińsko-Mazurskie region and Podlaskie region. These regions were selected as study sites because of their relative importance for the restructuring process in the dairy sector. Though one has to be cautious when generalising about the results obtained, it is reasonable to expect that regions that are less advanced in terms of restructuring would follow the path chosen by those two regions.

The paper is organised into five sections. The first section has set out the background. The second presents the main trends for milk production and milk consumption using a 'macro' perspective. The third takes a closer look at characteristics of households producing milk and changes that took place in the processing industry. The fourth discusses the results of econometric analyses assessing the main determinants of farm modernisation as well as the impact of dairy supply chain modernisation on households' incomes. Finally, the fifth section summarises the findings and draws main conclusions.

Milk production and consumption during transition

This section presents the main developments with regard to number of dairy producers, dairy cow herds, trends in marketed production, and demand for dairy products.

Milk production

Production trends

The main changes in milk production from 1989 to 2007 are shown in Table 1. Several interesting trends merit mention. During the first six years after market reform, the dairy herd declined by about 28% and milk yield per cow decreased by over 4%. These changes resulted in serious negative consequences for milk output. Milk production, adjusted to real demand, dropped by 28% to reach its lowest level of 11.3 m t in 1995 (GUS, various years). Since 1996, after the shock of the earlier period[5], the situation has stabilised. In that time the Agricultural Market Agency[6] openly intervened and purchased butter and skimmed milk powder, and introduced the first measures to protect the internal market. Real milk prices then started to increase, and so did milk production (IERiGŻ, 2005). During the EU pre-accession period, the restructuring of the dairy sector accelerated. The achievement of EU standards (especially sanitary and veterinary norms, and milk quality requirements), as well as the implementation of the Common Agricultural Policy (CAP) instruments (mainly preparations to implement the milk quota system), stimulated producers to start modernising their processes and to increase their scale of production. Investments, financed from farmers' own resources, loans granted by banks and dairy processing enterprises, and pre-accession support, resulted in an enormous improvement in milk quality. In the period 1999-2005, the share of extra-class milk (according to the EU standards) in total milk deliveries increased from 35% to 92%. For dairies with an EU certificate, this share was even higher, and accounted for 98% of milk deliveries (IERiGŻ, 2005). These strict quality requirements also brought negative social consequences, however. Many mainly small, inefficient producers were not able to adjust, and were thereby forced to either quit milk production or change to semi-subsistence farming. By 2005, even though there were 712,000 farms with dairy cows, only about 48% of them were delivering milk or milk products to the market (see Table 2). It is important to note also the relative absence of change in dairy farming intensity. The number of animals per 100 ha of agricultural land has fluctuated around 33-35 with no significant deviations from this value in the last decade (GUS, 2006).

---

[5] During 1990-95 period nominal milk purchase prices increased by about 11 fold while nominal farm input prices increased about 38 fold and the price of consumer goods increased 35 fold. Real milk price index only amounted to 32 per cent, which resulted in a significant decrease in the profitability of milk production (IERiGŻ, 2005, p.55).

[6] The Agricultural Market Agency is an intervention agency for agricultural markets. Its main role is to manage all the government's intervention measures.

Table 1. Characteristics of milk production in Poland from 1989-2005 (IERiGŻ, various volumes; GUS, various volumes).

	1989	1990	1994	1998	2000	2001	2002	2003	2004	2005	2006	2007
Number of dairy cows [1,000 heads]	4,994	4,919	3,863	3,471	3,098	3,005	2,873	2,897	2,796	2,795	2,824	2,787
Index %	100	98.5	77.4	69.5	62.0	60.2	57.5	58.0	56.0	56.0	56.5	55.8
Milk yields [litres/cow/year]	3,260	3,151	3,121	3,491	3,668	3,828	3,902	3,969	4,083	4,200	4,200	4,300
Index %	100	96.7	95.7	107.1	112.5	117.4	119.7	121.7	125.2	128.8	128.8	131.9
Milk production [million litres]	15,926	15,371	11,866	12,178	11,494	11,538	11,527	11,546	11,478	11,600	11,633	11,750
Index %	100	96.5	74.5	76.5	72.2	72.4	72.4	72.5	72.1	72.8	73.0	73.8
Milk deliveries [million litres]	11,385	9,829	6,269	7,070	6,583	7,025	7,219	7,316	7,997	8,831	8,419	8,380
Deliveries in total milk production %	71.5	63.9	52.8	58.1	57.3	60.9	63.2	63.4	69.7	76.1	72.4	70.9

Table 2. Number of dairy farms in 1990-2005 (IERiGŻ, various volumes; GUS, various volumes).

	1990	1996	2002	2003	2004	2005	2007
1.Number of producers [x1000]	1,831	1,309	876	810	735	712	657
Index %	100	71.5	47.8	44.2	40.1	38.8	35.9
2. Number of producers delivering to processing [x1000]	835	560	376	356	312	294	247
as % of total producers	45.6	42.8	42.9	44.0	42.5	41.3	37.6
3. Number of producers delivering directly to the market [x 1000]	n.a.	n.a.	n.a.	n.a.	76	50	27.5
as % of total producers	-	-	-	-	10.3	7.0	4.2

n.a.: Not assessed.

The average herd size in Poland is relatively small compared with West Europe. Nevertheless, a considerable increase has been observed. The average herd size increased from 2.7 in 1990 to 3.9 in 2005 (GUS, various volumes).

In 2005, the total number of cattle (beef and dairy cows) was 5.4 m head. Dairy cows account for roughly 50% of total cattle and this share has remained more or less stable during the last decade.

Inputs

Significant increases in input prices have been observed for all relevant commodities (i.e. fertilisers, pesticides, energy, machinery and seed grain). It is noteworthy that in the last decade (1995=100) the price scissors index (i.e. an index of price relationship of marketed agricultural products to goods and services purchased by private farms) has gradually decreased, and in 2005 reached the level of 69,4 (GUS, various volumes).

The respondents, asked during the qualitative research project, agreed unanimously that there have been enormous changes in technology and forage use during the past decade. One farmer said that, 'ten years ago I did not even think that I would ever have a 'western' tractor'. All our respondents had cooling tanks on their farms and larger producers had started to think about buying milking machines. As for forage, producers had switched from hay to hay-silage. Although it is more demanding in terms of machinery and thus involves higher expenditure, it is much easier to manage and certainly pays off. One of the respondents described this change as '[It is like] moving from a carriage to a car'. Green feed is mainly produced on the farm, and concentrates are bought in the market. The larger the producer, the more forage that is needed. As far as contact between farmers and input suppliers is concerned, for both technology and forage, farmers have no problem approaching a supplier.

It is also worth mentioning that sometimes the dairies negotiate with the input suppliers on behalf of the farmers. This certainly improves the farmers' position. Our respondents say that this is not very common, however. It is more common for contacts between farmers and input suppliers to take the form of cashless transactions. Input suppliers get their money from the dairy and farmers pay the dairies for their input supplies via milk sales. It seems that this type of arrangement has no effect on the power relationships between the parties involved.

Breeds and advisory services

The main breed in Poland is the Polish Holstein-Friesian. It accounts for roughly 97% of the total recorded cows[7]. Other breeds include Simmental, Red Polish, Jersey, White-back, Montbeliarde, Polish Black-White and Polish Red-White.

Dairy farms in Poland may take advantage of several sources of advice. For example, they may use advice provided by the Polish Federation of Cattle Breeders and Dairy Farmers. This organisation is authorised by the Ministry of Agriculture and Rural Development to keep herd books for dairy cattle. Advisory services to farmers are also provided by the Agricultural Advisory Centres which are located in every region and have local branches, and by almost every dairy processor.

Milk consumption

The transition period was characterised by a systematic decrease in milk consumption *per capita*. In 2004, the average milk consumption *per capita* amounted to 174 litres which was 67 litres (28%), lower than in 1990. This decrease could be attributed to the fact that a large number of households withdrew from milk production. It may be assumed that, particularly those households formerly

---

[7] In 2007 number of recorded cows amounted to 541,307.

producing milk mainly or only for subsistence purposes, considerably reduced their consumption of dairy products. Another probable reason for the decrease in milk consumption was the relative increase in dairy product prices when compared to prices of other food commodities (IERiGŻ, 2005). In future, milk consumption is cautiously and optimistically projected to increase slightly.

Consumption patterns

It is important to note that even though prices drive consumption decisions, the pattern of dairy product consumption has been changing. According to the director of a local supermarket chain in Warmińsko-Mazurskie region people are becoming more aware of what they eat and in the last few years demand for traditional dairy products (i.e. without artificial flavours) has increased by 20-30% per year.

All the people interviewed stated that there is a new trend in consumption patterns. Consumers are paying more attention to dairy product quality. One of the respondents said: 'before, consumers wanted to buy yellow cheese, now they want to buy, for example Gouda, and more often than not they want to buy Gouda from a given dairy processor'. Consumers also look for fresh products, and have started to avoid dairy products with distant expiry dates. In addition, regional products have become more popular (even in some super- and hyper-market chains). According to those interviewed, this could be a market outlet for small dairy processors.

It might be worth noting here that, including products made of goat and sheep milk, currently there are 37 traditional dairy products officially recognised by the Ministry of Agriculture and Rural Development.

Dairy farms and dairy processors

Dairy farms

As noted above, a significant decrease in number of households having cows was observed in Poland since the beginning of transition period. However, this was not the only remarkable feature characterising local dairy farms. Farms that decided to stay in milk production specialised and modernised, and in households having cows, dairy production dominated other agricultural activities. According to the experts interviewed, milk production has recently become the main agricultural activity in both surveyed regions. Although milk production has a long tradition, especially in Podlaskie, households tended to have diversified agricultural activities, including pig and sheep farming, and growing of tobacco. Today, local farmers specialise in milk production. In the opinion of our respondents this holds true for both small and large producers, but small-scale producers also need to combine earnings from farming with some other non-agricultural income, while large farmers can rely on agricultural income alone.

According to our respondents, there are several reasons behind the specialisation in milk production. First, the dairy industry is one of the few processing industries in these regions that survived the transformation to a market economy. Second, after shutting down rural collection points during 1990s, the outlets for agricultural products other than milk were significantly reduced. Third, farms with mixed production need access to machines for specific work in the field such as ploughing and harvesting. The collapse of communism resulted in the liquidation of many farmers' organisations that had provided such services, so today mixed production requires a substantial investment in machines, which only a few farmers can afford. Finally, the increase in both internal and external competition forced farmers to look for the most efficient use of their resources. Both these regions have relatively poor quality land more suited for grassland, and hence milk production, than to crop production.

Dairy processors

Improvements in the quality and range of the final products and the improvement in the quality of raw milk were the two main changes in the processing sector over the past 10 years. The need for concentration in the industry was also acknowledged and this has been particularly important during the past 5-6 years. The number of processors decreased in the last decade by about 30%, and in 2007 amounted to 232 dairies (IERiGŻ, 2005). Large dairy companies were often the initiators of this consolidation process. Furthermore, important changes occurred in the profitability of the sector at the time of, and in the period following, the accession process. After being somewhat modest in 1990s, profitability increased in the year of EU accession, as a result of increased export demand and a good Euro/Zloty exchange rate. However, in the following years the level of profitability progressively decreased due to increasing raw milk prices and an increase in production costs.

Increased competition in the dairy market forced dairy processors towards product specialisation. In large companies, this takes the form of dividing production among particular dairy plants, which allows them to provide consumers with a wide range of products to meet current demand. Smaller processors, with a less developed marketing system and poorer access to funds, have limited possibilities to introduce innovations, and thus cannot quickly adjust to consumer (or retail) requirements. Therefore, not being able to compete with larger companies in terms of size of deliveries or variety of products, smaller companies must find their own niche and produce unique products or products for further processing, such as skimmed milk powder (SMP). Some small processors specialise in exports.

In the opinion of both experts and segment representatives, the most significant factors influencing changes in the processing industry were: (1) the transition period, which influenced the situation mainly in the 1990s, (2) the pre-accession process, which started in 1998[8], and (3) integration into the EU. All these aspects required significant adjustments for institutions and policies. The necessary legal adjustments were also introduced, and support programmes for processors and producers were launched to assist them in meeting consumer quality requirements. Respondents also mentioned that the transformation of the retail sector (together with its internationalisation and consolidation) was an important factor. Retail expansion opened new outlets for dairy products, but at the same time imposed new requirement on dairies.

Difficulties and development constraints at the processing level are important, not only for dairies, but also for producers delivering raw products. For producers, the dairy plant is the most important segment of the market chain. It was observed that dairy processors not only play the role of milk purchaser, but also assist in farm development, for instance, by organising traineeships or providing short-term loans. In this context, dairy processors could be seen as one of the main drivers stimulating restructuring at the farm level. According to respondents, the main problems processors face when competing on the market can be divided into two groups: (1) barriers arising from the legal regulations, and (2) barriers related to the low economic efficiency of companies.

The milk quota system is the most frequently-mentioned legal barrier, both for the milk processors and for producers. The main problems with milk quotas are the production[9] ceiling and regional quota trading restrictions. The quota is allowed to be traded only between farmers who have their

---

[8] Despite the fact that the Polish pre-accession agreement was signed in 1994, the most significant arrangements concerning milk market regulation took effect at the end of 1990s. Therefore, restructuring the dairy sector became more important.

[9] Deliveries to processing plants during the first milk quota year (2004/2005) were 13% lower than quota assigned. However, due to rapid development of marketed milk production the milk quota in 2005/2006 was exceeded. In 2006/2007 and 2007/2008 milk quota was used in ca. 97%. It should be noted though that in the regions studied milk quota was exceeded and therefore the importance of this factor acting as a barrier to further restructuring is relatively higher than in other regions.

holding in the same administrative region. This results in high quota prices in dairy developing regions and inhibits the restructuring of milk production.

The other group of barriers relates to factors influencing the low economic efficiency of dairy processing. The main constraints here are:

- a weak position in the chain (except for the largest companies) being squeezed between producers' pressure (especially those with large deliveries) for high milk prices and retail pressure for cheap products;
- the poor management skills and low educational level of dairy employees (especially in dairy co-operatives);
- low labour efficiency, which negatively affects dairies' comparative advantage;
- the low level of consumption due low incomes resulting in people being unable to afford expensive dairy products. Smaller dairy processors especially have to compete on the local markets by lowering prices. This is extremely difficult when the processing margin is low.

Some experts have also said that a significant barrier to development of the processing sector is the form of co-operative ownership[10]. Unclear, disaggregated ownership rights result in more difficult management and a more complicated decision-making process. Some experts mentioned that the problem lies in the co-operative law, which hampers flexible management and restructuring of dairy co-operatives. This form of ownership dominates and in 2005 was responsible for 80% of processed milk and 70% of sales value (IERiGŻ, 2005).

Large-scale processors usually look to deliver their products to supermarkets/hypermarkets, where they can supply a large volume of produce. For those processors, large-scale retailers are also more stable partners than wholesalers, but they dictate stricter trade conditions and negotiate lower prices. The importance of the large retail channel has progressively increased since the end of 1990s, when the dominant form was wholesale, local cooperative chains and independent shops. However, according to our experts, only about 5% of milk produce is channelled through supermarkets and hypermarkets, although for fresh milk products (yoghurts), special cheeses and UHT milk, it reaches 20%. According to our research in both regions, large dairies sell from 30% to 60% of their products to large retail chains, 35-50% is channelled through the wholesale segment, and about 5% is delivered to local chains or independent shops. For small dairies as well as independent shops, the wholesale segment is dominant share of sales. Smaller dairies usually have an insufficient scale of production to deliver to supermarkets/hypermarkets.

It should be noted that the Polish dairy processing sector is exceptional compared to its counterparts in other countries in the region regarding the role played by foreign companies. In contrast to the situation commonly observed in other Central and Eastern European countries, dairy processing in Poland is dominated by domestic enterprises. The importance of foreign companies on the market is moderate though it needs to be stressed that their presence has indirectly contributed to the restructuring process by forcing domestic companies to maintain competitiveness (Dries and Swinnen, 2004).

Determinants and benefits of dairy farms' modernisation

Models

In order to further explore the issue of dairy farm modernisation, the following section highlights the main factors stimulating the restructuring process at farm level and assesses the benefits of modernisation. The results presented below are based on the findings of Milczarek-Andrzejewska *et al.* (2007).

---

[10] It should be noted however, that the largest dairy companies in Poland are domestic co-operatives.

One of the remarkable features of the dairy sector restructuring was the evolution of the modern marketing channel. Traditionally, milk was delivered by farmers to collection points and later collected by dairies. In the modern marketing channel, milk is collected directly from the cooling tank at the farm by a dairy truck. Obviously, joining the modern channel required considerable investments for modernisation at farm level (buying a cooling tank, expanding the herd size to increase scale of production and investments to become profitable). It is interesting therefore to investigate which factors were crucial in determining farms' capabilities to modernise and adjust to modern channel requirements. Second, it is of importance to check whether joining the modern marketing channel, and thus undertaking the modernisation effort, was beneficial to farmers. In order to do so, the income situation of modern channel farmers is compared with that of traditional channel farmers.

These comparative analyses are exposed to two kinds of problems. First, due to the fact that the phenomena under examination are likely to be interdependent, it is difficult to distinguish between cause and effect. To illustrate this, it can be argued that access to financial capital facilitates farm modernisation, but it can also be argued that farm modernisation may facilitate access to external funds through its positive impact on farm credit worthiness. Similarly, one may argue that farm modernisation positively affects farm revenues but it would be difficult to argue that the opposite does not also hold. To solve this 'interdependence problem', retrospective data are used to define explanatory variables. For example, following the decision to modernise after 2001, the farm income situation in 2006 is explained by past data (gathered for 2001). By doing this, cause and effect can be more easily distinguished.

The second problem refers to the so-called endogeneity problem. This stems from the fact that both explanatory and explained variables could be correlated with unobserved factors. In this case the use of standard econometric techniques such as ordinary least squares (OLS) is inadequate. To avoid potential problems, an instrumental variable approach was adopted (Angrist and Krueger, 2001). Three instruments were used namely distance to the closest dairy, distance to the closest collection point, and share of surveyed farms in given region[11] having a cooling tank.

The assessment of the determinants for joining a modern marketing channel was carried out using a simple probit model. The model assessing the impact of modernisation on farm revenues was composed of two steps. First, the probability of belonging to the modern channel was estimated. Second, it was then used in the model assessing determinants of farm revenues as an explanatory variable (MODERN). The results of the former model are presented in Table 3, column 1, whereas the results of the latter model (second step OLS) are presented in Table 3, column 2. In both models the same number of explanatory variables was used. Variables aimed at capturing the impact of incentives faced by farmers included access to unearned income, off-farm job and credit. They also include dummy variables for farms that experienced delays in payment from the dairy and for refusal of milk by the dairy due to poor quality. Farm characteristics included physical capital resources namely assets and other machinery specific to dairy production, herd size, annual milk yield per cow and land resources (owned and rented). Household characteristics included age and education of the household head, labour resources, dummies indicating farms cooperating with other households and/or farms delivering to cooperative, as well as variables approximating the farmer's attitude towards risk and his propensity to leave farming.[12] Finally, local shifters aimed at capturing

---

[11] Region considered here is Gmina, NUTS 5 according to the European Union classification.

[12] Risk variables took values from zero to two, with zero denoting risk averse farmers, one denoting risk neutral farmers and two denoting risk lovers. Values were assigned based on the following question: 'Provided that there are no costs of changing the dairy you are currently supplying, would you change it to supply another dairy offering 20% higher price, having no guarantee that this higher price will hold in the future?'. Farmers answering 'yes' were classified as risk lovers, those answering 'do not know' as risk neutral and those answering 'no' as risk averse. Propensity to leave farming, on the other hand, was a dichotomous variable equal to one if a farmer would leave agriculture having an opportunity to find off-farm employment with the same remuneration and zero otherwise.

the effect of farm's location, i.e. the region where it is located as well as the extent of development of dairy farms in its closer neighbourhood.

Results

The results presented draw on 323 observations, of which 218 belonged in 2006 to the modern, and 105 to the traditional, marketing channels. First, determinants of market channel choice are discussed, followed by a description of benefits of joining the modern marketing channel.

Determinants of joining a modern marketing channel

Several interesting insights can be obtained from the figures in Table 3, column 1. First, the positive and highly significant impact of CREDIT clearly shows that access to external funds appears to be decisive for keeping up with demands of the market and of dairy enterprises. The importance of external financial sources is further strengthened by non significant impact of OFF-FARM JOB and LABOUR. The latter observations could be indicative of two things. They may indicate that off-farm jobs are mainly low-skilled jobs, and thus provide only limited remuneration, or they could be indicative of inefficient use of labour resources in agriculture.

It is interesting to note that the market channel choice appears to be unaffected by initial physical nor land resources. One may argue therefore, that being relatively backward in terms of physical capital was not preventing farms from joining the modern channel. It needs to be noted though that the decision to shift to the modern channel was significantly dependent on having large herds and having cow breeds of better quality in terms of milk yields. Another result worth noting here is the negative and statistically significant effect of COOPERATION. This tends to indicate that potential costs of remaining in the traditional channel (e.g. lower price, higher risk of milk refusals, lower quality premiums) may be outweighed by benefits created by cooperation. This result is all the more interesting as it suggests that farmers' collaboration, commonly recommended as a tool having great potential for stimulating further restructuring, does not necessarily have these desired effects.

Effects of joining the modern channel

Estimates of the model assessing determinants of farm revenues are reported in Table 3, column 2. Most importantly, the results provide strong evidence that inclusion in the modern marketing channel contributes to considerable improvement in farm finances.

This is clearly indicated by the positive and statistically significant coefficient of the variable MODERN. This result concurs with theoretical predictions as well as with results of other studies examining restructuring of the agrifood supply chain (e.g. Swinnen *et al.*, 2006; White and Gorton, 2005). The effect of belonging to the modern channel holds regardless of herd size. Therefore, inclusion in the modern marketing channel appears to be beneficial not only for the largest farms, as suggested by the positive and statistically significant impact of HERD, but also for farms of medium and smaller size. However, the situation in this respect is slightly different for the smallest farmers, with fewer than five cows, for whom joining the modern channel has been on average slightly less beneficial.

Not surprisingly, farm revenues are positively correlated with milk yields and herd size. Interesting insights are provided from the analysis of REFUSAL and OFF-FARM JOB, both of which enter the estimated equation with negative signs. The former finding is as expected and reflects the fact that farms having problems with satisfying quality standards had only limited opportunities to grow. The latter observation, on the other hand, deserves more attention since it might appear counterintuitive. In all normal cases, access to off-farm job opportunities is expected to positively affect farm revenues.

Table 3. Factors affecting the probability of belonging to a modern marketing channel (MMC) and determinants of farm revenues.

	Probit with weights Dependent variable: Market channel choice in 2006 (1=modern, 0=traditional)	Ordinary Least Squares (OLS) with weights Dependent variable: Natural logarithm of farm revenues in 2006
Market channel choice		
MODERN		0.236***
Incentives		
UNEARNED INCOME 2001	-0.613**	-0.067
OFF-FARM JOB 2001	0.084	-0.089**
REFUSAL 2001	-0.143	-0.069*
DELAYS 2001	0.364	0.018
CREDIT 2001	1.925***	0.007
Farm size and assets		
ASSETS SPECIFIC 2001	-0.082	-0.017
ASSETS MACHINERY 2001	-0.011	0.021
HERD 2001	0.328***	0.012***
YIELDS 2001	0.001***	0.000**
FARM REVENUES 2001		0.498***
LAND OWNED 2001	-0.008	0.004
LAND LEASED 2001	0.039	-0.001
Household characteristics		
AGE 2006	-0.029	-0.002
EXPERIENCE 2006	0.032**	-0.002
EDUCATION	-0.273	0.021
LABOUR 2006	-0.046	-0.005
COOPERATION 2001	-1.121**	-0.096*
OWNERSHIP COOPERATIVE 2001	0.170	-0.017
RISK	0.497	0.035
LEAVE	-0.085	0.047
Local shifters		
NEIGHBOURS MAJORITY	-1.120**	0.223
NEIGHBOURS FEW	-1.095**	0.350*
PODLASKIE	0.227	0.133
Instrumental Variables		
DISTANCE_DAIRY_2006	-0.012	
DISTANCE_POINT_2006	0.339**	
COOLING_TANK_2001	3.414*	
Constant	-5.033*	4.873***
Observations	323	322
R-squared		0.805

***, **, * denote 1%, 5% and 10% significance level respectively.

Source: Authors' farm households survey, 2007.

However, it is reasonable to assume that the rural labour market in Poland is heavily biased towards agricultural employment. Under these circumstances, undertaking off-farm work might be an expression of seeking any available employment in order to make ends meet, rather than having a stable and rewarding job. This hypothesis finds support in the negative and statistically significant correlation between the level of 2001 farm revenues and access to off-farm activities.

Interestingly none of the variables aimed at capturing the effect of human capital appeared to be statistically significant.

The following aspects should also be mentioned regarding the effects of joining the modern channel. Comparing households always in the modern channel (MMC) with those supplying the traditional marketing channel (TMC) and those who made the switch from TMC to MMC at some point after 2001 (CHANGED) revealed several interesting facts (Table 4).

First, the average growth of agricultural revenue per capita (2001-2006) in case of MMC and CHANGED farmers was ca. 40% higher than that observed for TMC farmers. Similar differences were noticed with respect to growth rates of revenues obtained from milk sales. Remarkable also is the fact that the increase in herd size in the class CHANGED was almost three times greater than that observed for TMC (ca. 52% in CHANGED vs. ca. 18% in TMC). It is worth noting that, although milk yields per cow were growing at about the same rate in all groups, output per cow in MMC and CHANGED was ca. 1000 l higher than in TMC. These differences were also reflected in differences in the milk prices obtained by different groups. The average milk price increase in case of farmers who entered MMC was roughly 36% whereas in case of TMC farmers it was only 26%. The other striking difference between the groups relates to the level of specialisation in milk production. While the share of specialised farms decreased in TMC, it increased by over 50% in CHANGED, and by 17% in MMC. It can then be concluded that one of the most important outcomes of joining MMC is specialisation in milk production. However, this is mainly true for large farms.

Table 4. Farm development as affected by the marketing channel supplied.

	TMC			CHANGED			MMC		
	2001	2006	Increase %	2001	2006	Increase %	2001	2006	Increase %
1 Agricultural revenue per capita [PLN]	10,240	16,317	59.3	14,874	30,027	101.9	19,306	37,794	95.8
2 Milk sales revenue [PLN]	18,676	32,022	71.5	34,152	85,249	149.6	72,306	13,7218	89.8
3 Herd size [head]	7.9	9.3	17.7	11.1	16.9	52.3	17.9	23.3	30.2
4 Yields [l/cows]	3,480	3,901	12.1	4,272	4,844	13.4	4,676	5,252	12.3
5 Average milk price [PLN/100/ l]	70.3	88.4	25.7	73.3	99.4	35.6	80.7	100.1	24.0
6 Specialised in milk production[1] [%]	47.3	42.4	-10.4	44.0	66.9	52.0	74.9	87.6	17.0

[1] Milk revenue >60% of total.

Reported numbers are weighted averages.

Source: Authors' farm households survey 2007.

Only about 5% of all farms delivering to MMC and specialising in milk production had less than 10 cows in 2006.

Conclusions

In response to the dynamic and thorough transformations which have occurred, and still are taking place, in the Polish dairy sector, this paper has highlighted the main developments that were observed during the past decade. Moreover, it analysed determinants of the choice of marketing channel among local dairy farms and investigated the possible impacts of this choice on farmers' financial situation and behaviour. The main conclusions drawn from this analysis could be summarised as follows.

Restructuring of the Polish dairy sector was characterised by a market decrease in the number of dairy farms, a decrease in the number of dairy processing companies, and huge improvements in milk quality and efficiency of milk production. A remarkable phenomenon observed was the evolution of the modern marketing channel, through which milk is delivered from farms to dairy processors. Entering the modern marketing channel seems to be conditioned by exogenous rather than endogenous factors. Access to funds to pay for the necessary adjustments is the critical factor, rather than human capital or households' initial physical-capital resources. Given that farm financial capital resources are limited, the marketing-channel choice is crucially dependent on having access to external funds. Entering the modern marketing channel is facilitated by having herds of larger size and improved breeds. No systematic evidence for small farms being excluded from the modern channel was found, although it seems that the smallest farms (with fewer than five cows) are marginalised through having no access to external funds, either from a bank or from a dairy.

Joining the modern marketing channel positively affects the farm financial situation. This effect has been found for all farms regardless of their size. For the smallest ones, however, the impact is of lesser magnitude. In this context, further restructuring should be encouraged since it not only improves average farm welfare but also has potential to reduce the incidence of rural poverty. Since the necessary adjustments obviously require substantial investments, there is a need to broaden farm access to external funding. This is especially important for the very small producers. In this connection, rather than lump-sum transfers, microcredit programmes would provide farmers with appropriate incentives to use a loan efficiently, and should be facilitated. Changing the marketing channel also influences farm level of specialisation. Farms delivering to the modern channel tend to concentrate on milk production. However, these are mostly larger farms (having more than 10 cows). Small farmers and those who remained in the traditional channel tend to search for off-farm sources of income. Apparently, this strategy does not allow them to reach the level of revenues enjoyed by larger farms. Therefore, there is a strong need for development of non-agricultural income opportunities in rural areas in order to improve welfare of the smallest farms and to encourage and enable less efficient farmers to quit milk production.

Acknowledgements

This paper is based on the results obtained in the Regoverning Markets project. For more information see the website: www.regoverning.markets.org. Authors would like to thank Jerzy Wilkin, Liesbeth Dries, Csaba Csaki, Jikun Huang, Tom Reardon and all participants of the seminars in Warsaw for helpful comments and valuable suggestions.

References

Angrist, J.D. and Krueger, A.B., 2001. Instrumental variables and the search for identification: from supply and demand to natural experiments. Journal of Economic Perspectives 15: 69-85.

Dries, L. and Swinnen, J., 2004. Foreign direct investment, vertical integration, and local suppliers: Evidence from the Polish dairy sector. World Development 32: 1525-1544.

GUS, (various years). Statistical yearbooks.

GUS, 2006. Rolnictwo w 2005 r. (Agriculture in the year 2005), GUS, Warszawa.

IERiGŻ, (various volumes). Rynek mleka. Stan i perspektywy (Dairy market, Current state of affairs and prospects), IERiGŻ Warszawa.

IERiGŻ, 2005. Rozwój rynku mleczarskiego i zmiany jego funkcjonowania w latach 1990-2005 (Development of the dairy market and changes in its functioning during 1990-2005) IERiGŻ, Warsaw, 21/2005.

Milczarek-Andrzejewska, D., Malak-Rawlikowska, A., Fałkowski, J. and Wilkin, J., 2007. Farm level restructuring in Poland. Evidence from dairy sector. Regoverning Markets Agrifood Sector Study, IIED, London.

Swinnen, J.F.M., Dries, L., Noev, N. and Germeni, E., 2006. Foreign investments, supermarkets, and the restructuring of supply chains: Evidence from Eastern European dairy sectors. LICOS Discussion Papers, 165/2006.

White, J. and Gorton, M., 2005. A comparative study of agrifood chains in Moldova, Armenia, Georgia, Russia, and Ukraine. In: J.F.M. Swinnen (ed.), Case studies. The dynamics of vertical coordination in agrifood chains in Eastern Europe and Central Asia. World Bank Working Paper no. 42, pp. 5-43.

Wilkin, J., Milczarek, D., Malak-Rawlikowska, A. and Fałkowski, J., 2006. The dairy sector in Poland. Regoverning Markets Agrifood Sector Study, IIED, London.

Cattle sector and dairy chain developments in Kazakhstan, Kyrgyzstan and Uzbekistan

T. Karymsakov[1], A. Svitoys[2] and K. Elemesov[3]

[1]Scientific Industrial Center of Animal Production and Veterinary, Department of Dairy Cattle Breeding, O. Zandosov street 51, 050035 Almaty, Kazakhstan; kartalgat@yahoo.com; [2]Baltic Foundation, S. Konarskio str. 49, 03123, Vilnius, Lithuania; [3]Kazakhstan Association of Animal Production, Astana, Kazakhstan

Abstract

In the last 18 years the reforms in the agricultural sectors in the three countries are described. There were many basic changes that took place. In the years of economic transformation, the socialist agricultural enterprises raising scheduled breeds of cattle were transformed into various forms of private enterprises. This resulted in the dispersal of livestock to individuals, and only in those cases where herds are transferred collectively to private property was a high concentration of livestock maintained. In later years, with adaptation to the market economy and rigid competition between producers, some units gradually increased their numbers of livestock, improved the efficiency of management of the herds, and increased their profitability. The number of livestock gradually increased and will soon reach the same level as in the highest year of 1991. Total output of milk has already exceeded the earlier highest level with 10.23 t. The increase in production in recent years is connected with an increase in efficiency of cows in all categories of farms. The increase in production is also due to the use of semen from bulls, which are improved by foreign genetics, leading to an increase in milk yield per cow. This way, new breeds, types and lines of cattle are emerging. Following the privatisation of the state farms no effective breeding programme was retained. Up to 70-80% of the state farms have become outdated in mechanisation and the technical equipment is poor as is the quality of the forage reserves. In this context, in each country, state programs are carried out with the objective of (1) creating a station for artificial insemination in each area; (2) training young experts in animal breeding; (3) improving the pedigree structure of cattle in household facilities; (4) managing the breeding account on a personal computer; (5) modernising technical equipment and milking machines; (6) modernising factories for processing milk; (7) increasing finance for the support and development of agrarian and industrial enterprises; (8) decreasing the value added tax (VAT) on imported breeding cattle; (9) giving special equipment on lease; and (10) giving long-term credit.

Keywords: state programs, livestock population, milk production, animal breeding

Introduction

As is well known, some republics of Central Asia (Kazakhstan, Kyrgyzstan and Uzbekistan) were part of the Soviet Union where all agricultural concerns, including animal production had the state market system. In these three countries, as in all Socialist Republics, agriculture was operated within the state system in Kolkhoz and Sovkhoz. Both Kolkhozes and Sovkhozes were involved in breeding farm animals, including cattle, which were bred on the basis of scientific selection with a view to increasing production. This has led to the development of new domestic breeds and a different direction of production. In the middle of last century in the three republics, more specialised cattle breeds emerged from cross breeding local animals with more productive breeds. These included 2 dairy breeds, 3 dairy and meat breeds, and 5 breeds specialised for meat production.

In 1991, the population of cattle in these three republics reached 15.5 m, comprising 4.6 m in Uzbekistan, 1.2 m in Kyrgyzstan and 9.8 m in Kazakhstan. However, after independence the population of cattle sharply declined. This was connected with the reorganisation of all agricultural state systems and some kolkhozes and sovkhozes, because of transition from state ownership into private ownership. After this, herds with a high number of livestock and 85-90% of the industrial-technological cattle holdings were liquidated.

However, some animals were transferred to collective and co-operative farms where everyone who had the property, had been entrusted with one head. After this, these farms experienced a very complex process of transition to a market system, resulting in a rise in population of cattle and an increase in cattle production. A summary of the changes in cattle numbers and milk production is shown in Table 1.

The trend towards an increase in the cattle population and the total output from cattle in the three countries was first observed at the end of 20th century. In each republic, raising of cattle was based on a certain conception and increasing economic expectations.

Table 1. Cattle numbers and milk production for 1991, 2000 and 2007 in Kazakhstanaн, Uzbekistan and Kyrgyzstan.

	1991	2000	2007
Population of cattle (m)	15.53	10.47	13.89
Milk production (m t)	9.73	8.36	10.75

Kyrgyzstan

Cattle raising in the Republic of Kyrgyzstan is one of the main branches of agriculture It represents 60% of gross output of the animal industries. Raising of cattle is favoured by good climatic conditions and social factors. It occurs on 83% of farmland, of which 9.6 million ha is in natural pastures. Cattle breeding is one of the main branches of animal production. In the republic, two breeds of cattle namely Alatau and Aulietinsc exist, and in the high mountains yaks predominate.

The Alatau breed represents the dairy-meat type and was approved in 1950. In high mountain conditions, this breed surpasses the local Kirghiz cattle on dairy and meat production. Nowadays, it is raised in all regions of the republic.

The Aulietinsc breed is a dairy breed and was established as an independent breed in 1974. It was formed by crossing local cattle with bulls of a Dutch breed.

Yaks are raised in high mountainous areas which are unsuited to other kinds of farm animals. The local population has been engaged in the raising of yaks for a long time, but recently the numbers of yaks has decreased.

After independence, the number of yaks decreased by 39,000 in the period 1991 to 2000. However, since 2001 the population has gradually increased again, and in 2007 it had grown by 6,100 to a total of 22,400 animals (Figure 1).

As can be seen in Figure 1, the population of cattle declined by 3.14 m from 1991 to 1996. From 1997 on, a tendency towards an increase was observed and by 2007, the population of cattle was 1.21 m head.

Milk production fell to 251,400 t in the period 1991 to 1996, but from 1997 to 2000, milk volume increased by 169,000 t and by then had reached 1.06 mt. From 2001 to 2007, volume increased by 149,500 t and has now reached 1.21 m (Figure 2). This means that the country has extra production in comparison with 1991 by 42,000 t. The increase of milk production has occurred, not from increases

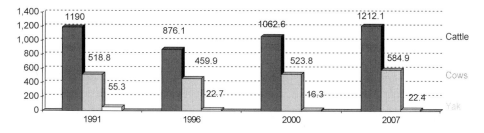

Figure 1. Change in population of cattle from 1991 to 2007 (×1000 head).

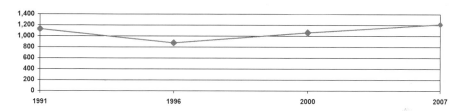

Figure 2. Change in the milk production volume from 1991 to 2007 (t).

in production per cow, but because of an increase in the population of cows. The average milk yield per cow remains rather low at 2,118 kg.

There are more than 183 farms involved in the production of farm animals, including 65 breeding farms for cattle and 7 for yak. At the present stage of breeding development the main task is improving existing breeds. There are two state breeding farms for the Alatau breed and four farmer breeding enterprises for the Alatau and Holstein-Friesian breeds (Figure 3).

As can be seen from Figure 3, the distribution of breeding enterprises shows that the main operators prefer breeding of highly productive cattle for milk and meat production.

Now, cattle breeding is improving with the assistance of work carried out by the Central Asian breeding service, LTD, which was created with support from the Kyrgyz-Swedish agriculture project 'Helvitos'. Today, 30 cattle breeding facilities are operating a program of genetic improvement of quality of cattle and all are provided with semen from Swedish companies. In the long term, the development of the animal industries should focus especially on meat and dairy cattle breeding.

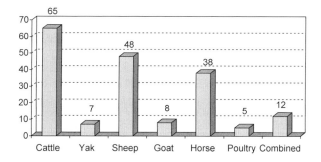

Figure 3. Number of management enterprises for various farm animal species.

Uzbekistan

In the Republic of Uzbekistan, on the mountain and foothill zones, there is enough pasture for profitable rearing of cattle as it does not incur any costs other than pasture. In these zones there are many different breeds and systems of cattle production: Black-Motley, Red Steppe, Swiss, Bushu, Santa-Gertruda, as well as low-producing local breeds.

From 1991 to 1995, as compared with other former socialist countries, the population of cattle did not decrease, but in fact increased (Figure 4).

From 1990 to 1995, the cattle population increased by 920,000, but from 1996 to 2000, it declined by 200,000. This was associated with low productivity and inadequate forage reserves. From 2001 onwards, the trend was for an increase of population each year until 2007 when the population of cattle was 6.82 m. Such a high rate of growth in cattle numbers was associated with the attention given to the development of personnel, and farms were a major factor in the growth of employment for the population. There were increases in incomes and as whole in the standard of living, resulting in a steady saturation of home market by vital food stuffs such as meat, milk and many others.

Special attention in this context was given to stimulating increases in the population of cattle. It is legislatively determined, that people who raise cattle in a personal capacity, and on farms, qualify for a work-record card and a pension.

For the purchase of highly productive cattle by the population and leading farm enterprises there was an expansion of credit facilities by banks, and 'Fund of assistance of employment' is accepted. Realisation of the accepted measures will lead to a livestock population in 2010 of 8.6 m, resulting in a considerable increase in the number of the persons involved in cattle production. This will enhance the level of their material well-being.

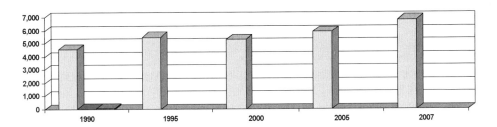

Figure 4. Change in population of cattle from 1990 to 2007.

Kazakhstan

The Republic of Kazakhstan has huge territory, different natural landscapes and different climates. The republic is rich in steppes, dense woods, mountain tops, deserts and semi deserts. In this context local cattle breeds are raised which are adapted to the natural and climatic conditions.

There are 5 mainly meat producing breeds and 4 milk plus meat producing breeds. The meat breeds are: Auliakol, Kazakh-wait-head, Santa-Gertruda, Kalmik and Galloway. The milk plus meat breeds are: Alatau, Simmental, Red Steppe and Auliakol.

After independence the cattle population decreased sharply, but since 2001 an increase in livestock numbers has been observed (Figure 5). From 1991 to 2000, livestock numbers fell from 9.76 m to 4.11 m. However, since 2001 there has been an increase in the population of livestock, and from 2001 to 2007 it increased by 982,000. Now the cattle population is 5.85 m.

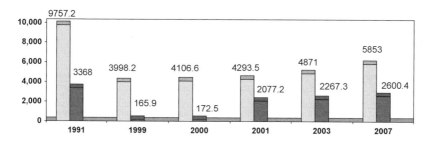

Figure 5. Population of cattle in Kazakhstan 1991 to 2007 (×1000).

The same tendency was observed in milk production. From 1991 to 2000, total milk production decreased from 5.6 m to 3.7 m (i.e. by 1.9 m). Milk yield per cow during this period decreased by 19 kg to 1,988 kg but since 2000, it has increased by 233 kg to 2,202 kg.

Today the country has 461 economic enterprises that raise farm animals of many species (Figure 6) but the population prefers to be engaged in cattle production, mainly dairying. Of the 166 enterprises engaged in cattle production, 109 are engaged dairying (Figure 7).

Of the 109 dairy cattle enterprises, 16 have breeding enterprise status which entitles them to state support for breeding animals, buying breeding bulls, and development of breeding stations.

Today, a program of improvement for increasing production of milk-meat breeds is ongoing. By using frozen semen from Holstein bulls, work is progressing on producing a new type of Black-Motley cattle of Kazakhstan. In the northern region, crossing of local Red Steppe and the Red-Danish cattle breeds is creating a new type of Red Steppe cattle. In the east and in the northern part of the country,

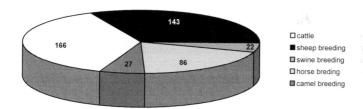

Figure 6. Number of economic enterprises raising farm animals in Kazakhstan.

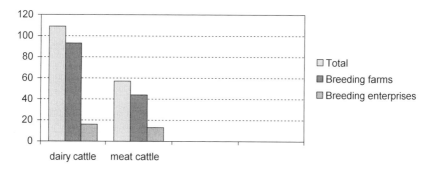

Figure 7. Number of farms which rear dairy and meat cattle.

work is at an advanced stage on producing a new Red-Motley type, based on crossing Simmental and the Holstein-Red-Motley breed. In the south-east based on the local Alatau breed and Swiss bulls of American origin, a new type of Kazakhstan brown dairy cattle 'Akirys' has already been produced. Such types of dairy cattle differ from their predecessors in that they have better dairy type, a more desirable shape of udder, and milk production that exceeds their predecessors by 1,500-2,000 kg. The population of breeding cattle (according to 2007 statistics) is 253,500, or 4.1% of total. Work is now in progress to increase the population of breeding livestock and improve the pedigree structure with the objective of increasing milk production. In this context, in 2007, 1,700 cows were imported from Canada. In 2008, the number of breeding enterprises is expected to increase to 500 units.

Part 3 Concluding remarks

Remarks and recommendations of the workshop

M. Zjalic

EAAP, Via G. Tomassetti 3, 00161 Rome, Italy

Cattle sector in Central Europe

In reviewing the situation and trends in the development of cattle production in Central European transition countries, the workshop noted that the process of transition of the cattle sector from centrally-planned to market economy system was characterised by a sharp decline in production and consumption, as well as in the number of cattle. These developments have been accompanied by changes in farm structures as a consequence of privatisation and de-collectivisation. The end of the last century was marked by further changes in farm structure, herd size and productivity: the number of dairy farms and dairy cows continued to decline, while herd size and per cow milk production are increasing. The trend of specialisation of cattle production – separation of beef from dairy production and establishment of beef farms, introduction of new beef breeds and use of dairy herds for beef production – although not evenly marked in all countries, is expected to continue. The system of milk quotas and the adjustment of agrarian policies (through the Common Agricultural Policy), which started even before the accession of these countries to the European Union, have imposed the need for a new type of farm management. Technology development and increase in farm size contributed to the establishment of closer links between milk producers and the milk processing industry and to the gradual integration of primary production and processing. Horizontal integration – establishment of cooperatives and producers' associations – is developing relatively slowly, because small farmers are reluctant to join cooperatives after the negative experience they had under the previous regime. In some countries, producers' associations play an important role as representative organisations and partners of industry and administration.

Participants at the workshop underlined the need for regular and continuous monitoring of the structural changes in the sector. Adjustments in policies should take into account market demand and trends as well as the social role of agriculture, particularly in view of the large number of small subsistence farms in some countries. Many of these subsistence farms have a potential to become market-oriented producers, while a certain number of them will abandon cattle production. Those who will remain in production need support and training in modern technologies and management. Participants considered that the system of milk quotas has contributed to the survival of milk production in the regions with the lower productivity of land and animals. The announced gradual abolishment of milk quotas (soft landing) was met with some understanding as a measure that would contribute to the better adjustment of the dairy sector to market demand. On the other hand, the specific position of the sector in Central Europe and its lower economic power and productivity should be taken into account in designing future policy measures. It was agreed that partners within the Cattle Network should continue to monitor policy developments of the CAP affecting the dairy sector, so as to be in the position to contribute through competent bodies to the assessment of the situation and to the formulation of new proposals, when required.

Noting that dairy herds are the basis for beef production in Central and Eastern Europe, as regular suppliers of calves for fattening in other Member States, participants agreed on the need to monitor the impact of the decline in the number of dairy cows on EU beef production.

The genetic improvement of dairy herds in Central Europe greatly contributed to the consolidation of the dairy sector and to a significant increase in productivity. The future breeding objectives should take into account all traits relevant to the efficiency and profitability of milk production and sustainability of the herd, looking also at traits related to fertility, longevity and animal health. The genetic potential of local breeds should be increased particularly in the production of local products and supply of niche markets, including orientation towards organic production.

Cattle sector in Eastern Europe

After the initial decline in number and productivity, the present situation of dairy cattle in the Russian Federation has become more stable and predictable than it was in the 1990s. The total milk production has been stabilised while the average milk yield per cow is constantly increasing, with a potential for future growth. The official policy document 'Concept-Forecast for Russia's Animal Husbandry Development up to Year 2010' envisages the following actions:
- Restoration and development of the population and structure as well as preservation of the unique gene stock of breeding animals.
- Creation of favorable conditions for investment policy in this sector.
- Raising the economic efficiency of activities pursued by breeding organisations and enterprises.
To resolve these issues, it has been decided first of all to improve the normative-methodological as well as economical and material foundation of cattle breeding, which will be aimed at:
- Increasing the number of breeding herds and animals under registration (identification, maintaining the data base to be used as a basis for official herd books of pedigree animals).
- Increasing the pace of genetic progress for breeding animal populations according to the selected characteristics due to the introduction and optimisation of breeding programs with the populations of farm animals.
- Optimisation of the breeding organisations infrastructure (breed associations, systems for farm animals artificial insemination, independent laboratories to register phenotypical characteristics and estimation of the animals' genetic value).
- Increasing the effectiveness of distribution of the best genetic resources, its rational use and realisation of the potential under the real conditions of agricultural production.
- Conducting the objective monitoring of the breeding livestock sector, projection of its development and optimisation of breeding programs.
- Russia's accession to international organisations dealing with pedigree animal husbandry.
Implementation of these steps into the practice of dairy cattle husbandry in the Russian Federation offers the opportunity to look ahead to the future with definite optimism.

In other former USSR countries, the dairy sector accounts for up to 25% of total agricultural production. Animal production in these countries suffered seriously during the transition period in the 1990's. For instance, in Ukraine milking livestock have decreased to less than half in 15 years, and by the end of 2005 amounted to only 45% of the 1990 number. After that rapid decrease in the cattle population and the corresponding decrease in milk production, a gradual improvement in the situation has since been observed. Some countries, like Belarus, have stabilised the number of cows, attained sustainable growth in yields and total milk production including a significant export of milk and milk products.

In other countries, the most important problems and constraints in dairy sector development include the prevalence of small-scale farms in the total raw milk supply often resulting in the production of low quality raw milk, constraints to accessing credit, low prices for milk, lack of investment in dairy farming, and underdeveloped logistics and infrastructure such as milk collection, storing and distribution. The production of feeds and fodders has decreased significantly and pastures are not well managed. The extent of artificial insemination use has sharply declined as centralised breeding farms have been abandoned and the core breeding stock have been distributed to private individuals, who are often not experienced in livestock breeding. This has led to the deterioration of the genetic characteristics of cattle. In addition, the high prevalence of zoonotic and transboundary animal diseases, such as brucellosis, tuberculosis, foot and mouth disease, hinder the development of the dairy sector in some countries. The current externally funded projects in Caucasian countries (e.g. several small technical projects in Georgia, FAO project on strategy for livestock development in Armenia) indicate the main lines of future policies and developments: establishment of an institutional and legal framework for animal production, establishment of technical capacities for genetic improvement, animal feeding, meat and milk processing, and trade in animals and animal products.

Appendix. Short history of the EAAP Cattle Network Working Group

In 2003, the EAAP Council established its Working Group 'Business Support Information Network for Cattle Sector' (Cattle Network) to facilitate business and technical co-operation among cattle breeders and producers from its Member Organisations. The Working Group was established after the successful completion of the BABROC[13] project funded by the European Commission and implemented by EAAP and its members from the Central and Eastern European countries – former candidates for the EU membership. The basic scope of the Cattle Network is the exchange of information to promote and facilitate business operations among breeders' and producers' associations. The Network monitors and discusses trends in production and consumption of cattle products and policy measures affecting the cattle sector. In doing so, it relies on studies and analyses produced by existing institutions and organisations. Activities of the Network are complementary to the existing representative structures at EU level (COPA/COGECA), the existing European and international scientific and technical entities, such as EAAP, ICAR/INTERBULL and the European breed associations. The Cattle Network has established its website http://www.cattlenetwork.net as a virtual gateway offering a complete range of both market- and consumer-oriented information as well as modern communication tools. It has already become an online meeting point for researchers, professionals, producers and consumers in the cattle sector.

Meetings and workshops of the Network are held on the occasion of the EAAP Annual meetings and on the occasion of other important events, such as international fairs and important cattle shows. Since 2003, the Network has organized two meetings and four workshops:

- 2003 – 1st meeting in Rome: Establishment of the Cattle Network Working Group.
- 2004 – 2nd meeting in Paris: Work plan for 2004-2005.
- 2005 – 1st Cattle Network Workshop 'Perspective of beef production in Europe', Sweden.
- 2006 – 2nd Cattle Network Workshop 'Development trend in small cattle farms', Turkey.
- 2007 – 3rd Cattle Network Workshop 'Adaptation and conformation of EU beef systems to CAP regulations', Ireland.
- 2008 – 4th Cattle Network Workshop 'Cattle sector development in transition countries of CEE', Lithuania.

---

[13] 'Support to the restructuring and strengthening of cattle producers' and breeders' associations as business representative organisations in Central and Easter European countries'